Environmental Issues

CLIMATE CHANGE

Environmental Issues

AIR QUALITY
CLIMATE CHANGE
CONSERVATION
ENVIRONMENTAL POLICY
WATER POLLUTION
WILDLIFE PROTECTION

Environmental Issues

CLIMATE CHANGE

Yael Calhoun
Series Editor

Foreword by David Seideman,
Editor-in-Chief, *Audubon* Magazine

CHELSEA HOUSE
PUBLISHERS
A Haights Cross Communications Company ®
Philadelphia

CHELSEA HOUSE PUBLISHERS

VP, NEW PRODUCT DEVELOPMENT Sally Cheney
DIRECTOR OF PRODUCTION Kim Shinners
CREATIVE MANAGER Takeshi Takahashi
MANUFACTURING MANAGER Diann Grasse

Staff for CLIMATE CHANGE

EXECUTIVE EDITOR Tara Koellhoffer
EDITORIAL ASSISTANT Kuorkor Dzani
PRODUCTION EDITOR Noelle Nardone
PHOTO EDITOR Sarah Bloom
SERIES AND COVER DESIGNER Keith Trego
LAYOUT 21st Century Publishing and Communications, Inc.

A Haights Cross Communications ✈ Company ®

First Printing

9 8 7 6 5 4 3 2 1

Library of Congress Cataloging-in-Publication Data

Climate change/[edited by Yael Calhoun]; foreword by David Seideman.
  p. cm.—(Environmental issues)
    Includes bibliographical references and index.
  ISBN 0-7910-8206-7
  1. Climatic changes—Juvenile literature. 2. Climatic changes—
Environmental aspects. I. Calhoun, Yael. II. Series.
QC981.8.C5C51129 2005
363.738'74—dc22

                                        2004028993

Contents Overview

Detailed Table of Contents

FOREWORD

by David Seideman, Editor-in-Chief, *Audubon* Magazine

For anyone contemplating the Earth's fate, there's probably no more instructive case study than the Florida Everglades. When European explorers first arrived there in the mid-1800s, they discovered a lush, tropical wilderness with dense sawgrass, marshes, mangrove forests, lakes, and tree islands. By the early 20th century, developers and politicians had begun building a series of canals and dikes to siphon off the region's water. They succeeded in creating an agricultural and real estate boom, and to some degree, they offset floods and droughts. But the ecological cost was exorbitant. Today, half of the Everglades' wetlands have been lost, its water is polluted by runoff from farms, and much of its wildlife, including Florida panthers and many wading birds such as wood storks, are hanging on by a thread.

Yet there has been a renewed sense of hope in the Everglades since 2001, when the state of Florida and the federal government approved a comprehensive $7.8 billion restoration plan, the biggest recovery of its kind in history. During the next four decades, ecologists and engineers will work to undo years of ecological damage by redirecting water back into the Everglades' dried-up marshes. "The Everglades are a test," says Joe Podger, an environmentalist. "If we pass, we get to keep the planet."

In fact, as this comprehensive series on environmental issues shows, humankind faces a host of tests that will determine whether we get to keep the planet. The world's crises—air and water pollution, the extinction of species, and climate change—are worsening by the day. The solutions—and there are many practical ones—all demand an extreme sense of urgency. E. O. Wilson, the noted Harvard zoologist, contends that "the world environment is changing so fast that there is a window of opportunity that will close in as little time as the next two or three decades." While Wilson's main concern is the rapid loss of biodiversity, he could have just as easily been discussing climate change or wetlands destruction.

The Earth is suffering the most massive extinction of species since the die-off of dinosaurs 65 million years ago. "If we continue at the current rate of deforestation and destruction of major ecosystems like rain forests and coral reefs, where most of the biodiversity is concentrated," Wilson says, "we will surely lose more than half of all the species of plants and animals on Earth by the end of the 21st century."

Many conservationists still mourn the loss of the passenger pigeon, which, as recently as the late 1800s, flew in miles-long flocks so dense they blocked the sun, turning noontime into nighttime. By 1914, target shooters and market hunters had reduced the species to a single individual, Martha, who lived at the Cincinnati Zoo until, as Peter Matthiessen wrote in *Wildlife in America,* "she blinked for the last time." Despite U.S. laws in place to avert other species from going the way of the passenger pigeon, the latest news is still alarming. In its 2004 State of the Birds report, Audubon noted that 70% of grassland bird species and 36% of shrubland bird species are suffering significant declines. Like the proverbial canary in the coalmine, birds serve as indicators, sounding the alarm about impending threats to environmental and human health.

Besides being an unmitigated moral tragedy, the disappearance of species has profound practical implications. Ninety percent of the world's food production now comes from about a dozen species of plants and eight species of livestock. Geneticists rely on wild populations to replenish varieties of domestic corn, wheat, and other crops, and to boost yields and resistance to disease. "Nature is a natural pharmacopoeia, and new drugs and medicines are being discovered in the wild all the time," wrote Niles Eldredge of the American Museum of Natural History, a noted author on the subject of extinction. "Aspirin comes from the bark of willow trees. Penicillin comes from a mold, a type of fungus." Furthermore, having a wide array of plants and animals improves a region's capacity to cleanse water, enrich soil, maintain stable climates, and produce the oxygen we breathe.

Today, the quality of the air we breathe and the water we drink does not augur well for our future health and well-being. Many people assume that the passage of the Clean Air Act in 1970 ushered in a new age. But the American Lung Association reports that 159 million Americans—55% of the population—are exposed to "unhealthy levels of air pollution." Meanwhile, the American Heart Association warns of a direct link between exposure to air pollution and heart disease and strokes. While it's true that U.S. waters are cleaner than they were three decades ago, data from the Environmental Protection Agency (EPA) shows that almost half of U.S. coastal waters fail to meet water-quality standards because they cannot support fishing or swimming. Each year, contaminated tap water makes as many as 7 million Americans sick. The chief cause is "non-point pollution," runoff that includes fertilizers and pesticides from farms and backyards as well as oil and chemical spills. On a global level, more than a billion people lack access to clean water; according to the United Nations, five times that number die each year from malaria and other illnesses associated with unsafe water.

Of all the Earth's critical environmental problems, one trumps the rest: climate change. Carol Browner, the EPA's chief from 1993 through 2001 (the longest term in the agency's history), calls climate change "the greatest environmental health problem the world has ever seen." Industry and people are spewing carbon dioxide from smokestacks and the tailpipes of their cars into the atmosphere, where a buildup of gases, acting like the glass in a greenhouse, traps the sun's heat. The 1990s was the warmest decade in more than a century, and 1998 saw the highest global temperatures ever. In an article about global climate change in the December 2003 issue of *Audubon*, David Malakoff wrote, "Among the possible consequences: rising sea levels that cause coastal communities to sink beneath the waves like a modern Atlantis, crop failures of biblical proportions, and once-rare killer storms that start to appear with alarming regularity."

Yet for all the doom and gloom, scientists and environmentalists hold out hope. When Russia recently ratified the Kyoto Protocol, it meant that virtually all of the world's industrialized nations—the United States, which has refused to sign, is a notable exception—have committed to cutting greenhouse gases. As Kyoto and other international agreements go into effect, a market is developing for cap-and-trade systems for carbon dioxide. In this country, two dozen big corporations, including British Petroleum, are cutting emissions. At least 28 American states have adopted their own policies. California, for example, has passed global warming legislation aimed at curbing emissions from new cars. Governor Arnold Schwarzenegger has also backed regulations requiring automakers to slash the amount of greenhouse gases they cause by up to 30% by 2016, setting a precedent for other states.

As Washington pushes a business-friendly agenda, states are filling in the policy vacuum in other areas, as well. California and New York are developing laws to preserve wetlands, which filter pollutants, prevent floods, and provide habitat for endangered wildlife.

By taking matters into their own hands, states and foreign countries will ultimately force Washington's. What industry especially abhors is a crazy quilt of varying rules. After all, it makes little sense for a company to invest a billion dollars in a power plant only to find out later that it has to spend even more to comply with another state's stricter emissions standards. Ford chairman and chief executive William Ford has lashed out at the states' "patchwork" approach because he and "other manufacturers will have a hard time responding." Further, he wrote in a letter to his company's top managers, "the prospect of 50 different requirements in 50 different states would be nothing short of chaos." The type of fears Ford expresses are precisely the reason federal laws protecting clean air and water came into being.

Governments must take the lead, but ecologically conscious consumers wield enormous influence, too. Over the past four decades, the annual use of pesticides has more than doubled, from 215 million pounds to 511 million pounds. Each year, these poisons cause $10 billion worth of damage to the environment and kill 72 million birds. The good news is that the demand for organic products is revolutionizing agriculture, in part by creating a market for natural alternatives for pest control. Some industry experts predict that by 2007 the organic industry will almost quadruple, to more than $30 billion.

E. O. Wilson touts "shade-grown" coffee as one of many "personal habitats that, if moderated only in this country, could contribute significantly to saving endangered species." In the mountains of Mexico and Central America, coffee grown beneath a dense forest canopy rather than in cleared fields helps provide refuge for dozens of wintering North American migratory bird species, from western tanagers to Baltimore orioles.

With conservation such a huge part of Americans' daily routine, recycling has become as ingrained a civic duty as obeying traffic lights. Californians, for their part, have cut their energy consumption by 10% each year since the state's 2001 energy crisis. "Poll after poll shows that about two-thirds of the American public—Democrat and Republican, urban and rural—consider environmental progress crucial," writes Carl Pope, director of the Sierra Club, in his recent book, *Strategic Ignorance*. "Clean air, clean water, wilderness preservation— these are such bedrock values that many polling respondents find it hard to believe that any politician would oppose them."

Terrorism and the economy clearly dwarfed all other issues in the 2004 presidential election. Even so, voters approved 120 out of 161 state and local conservation funding measures nationwide, worth a total of $3.25 billion. Anti-environment votes in the U.S. Congress and proposals floated by the like-minded Bush administration should not obscure the salient fact that so far there have been no changes to the major environmental laws. The potential for political fallout is too great.

The United States' legacy of preserving its natural heritage is the envy of the world. Our national park system alone draws more than 300 million visitors each year. Less well known is the 103-year-old national wildlife refuge system you'll learn about in this series. Its unique mission is to safeguard the nation's wild animals and plants on 540 refuges, protecting 700 species of birds and an equal number of other vertebrates; 282 of these species are either threatened or endangered. One of the many species particularly dependent on the invaluable habitat refuges afford is the bald eagle. Such safe havens, combined with the banning of the insecticide DDT and enforcement of the

Endangered Species Act, have led to the bald eagle's remarkable recovery, from a low of 500 breeding pairs in 1963 to 7,600 today. In fact, this bird, the national symbol of the United States, is about be removed from the endangered species list and downgraded to a less threatened status under the CITES, the Convention on International Trade in Endangered Species.

This vital treaty, upheld by the United States and 165 other participating nations (and detailed in this series), underscores the worldwide will to safeguard much of the Earth's magnificent wildlife. Since going into effect in 1975, CITES has helped enact plans to save tigers, chimpanzees, and African elephants. These species and many others continue to face dire threats from everything from poaching to deforestation. At the same time, political progress is still being made. Organizations like the World Wildlife Fund work tirelessly to save these species from extinction because so many millions of people care. China, for example, the most populous nation on Earth, is so concerned about its giant pandas that it has implemented an ambitious captive breeding program. That program's success, along with government measures prohibiting logging throughout the panda's range, may actually enable the remaining population of 1,600 pandas to hold its own—and perhaps grow. "For the People's Republic of China, pressure intensified as its internationally popular icon edged closer to extinction," wrote Gerry Ellis in a recent issue of *National Wildlife*. "The giant panda was not only a poster child for endangered species, it was a symbol of our willingness to ensure nature's place on Earth."

Whether people take a spiritual path to conservation or a pragmatic one, they ultimately arrive at the same destination. The sight of a bald eagle soaring across the horizon reassures us about nature's resilience, even as the clean air and water we both need to survive becomes less of a certainty. "The conservation of our natural resources and their proper use constitute the fundamental problem which underlies almost every other problem of our national life," President Theodore Roosevelt told Congress at the dawn of the conservation movement a century ago. His words ring truer today than ever.

Introduction: "Why Should We Care?"

Our nation's air and water are cleaner today than they were 30 years ago. After a century of filling and destroying over half of our wetlands, we now protect many of them. But the Earth is getting warmer, habitats are being lost to development and logging, and humans are using more water than ever before. Increased use of water can leave rivers, lakes, and wetlands without enough water to support the native plant and animal life. Such changes are causing plants and animals to go extinct at an increased rate. It is no longer a question of losing just the dodo birds or the passenger pigeons, argues David Quammen, author of *Song of the Dodo*: "Within a few decades, if present trends continue, we'll be losing *a lot* of everything."[1]

In the 1980s, E. O. Wilson, a Harvard biologist and Pulitzer Prize–winning author, helped bring the term *biodiversity* into public discussions about conservation. *Biodiversity*, short for "biological diversity," refers to the levels of organization for living things. Living organisms are divided and categorized into ecosystems (such as rain forests or oceans), by species (such as mountain gorillas), and by genetics (the genes responsible for inherited traits).

Wilson has predicted that if we continue to destroy habitats and pollute the Earth at the current rate, in 50 years, we could lose 30 to 50% of the planet's species to extinction. In his 1992 book, *The Diversity of Life*, Wilson asks: "Why should we care?"[2] His long list of answers to this question includes: the potential loss of vast amounts of scientific information that would enable the development of new crops, products, and medicines and the potential loss of the vast economic and environmental benefits of healthy ecosystems. He argues that since we have only a vague idea (even with our advanced scientific methods) of how ecosystems really work, it would be "reckless" to suppose that destroying species indefinitely will not threaten us all in ways we may not even understand.

THE BOOKS IN THE SERIES

In looking at environmental issues, it quickly becomes clear that, as naturalist John Muir once said, "When we try to pick out anything by itself, we find it hitched to everything else in the Universe."[3] For example, air pollution in one state or in one country can affect not only air quality in another place, but also land and water quality. Soil particles from degraded African lands can blow across the ocean and cause damage to far-off coral reefs.

The six books in this series address a variety of environmental issues: conservation, wildlife protection, water pollution, air quality, climate change, and environmental policy. None of these can be viewed as a separate issue. Air quality impacts climate change, wildlife, and water quality. Conservation initiatives directly affect water and air quality, climate change, and wildlife protection. Endangered species are touched by each of these issues. And finally, environmental policy issues serve as important tools in addressing all the other environmental problems that face us.

You can use the burning of coal as an example to look at how a single activity directly "hitches" to a variety of environmental issues. Humans have been burning coal as a fuel for hundreds of years. The mining of coal can leave the land stripped of vegetation, which erodes the soil. Soil erosion contributes to particulates in the air and water quality problems. Mining coal can also leave piles of acidic tailings that degrade habitats and pollute water. Burning any fossil fuel—coal, gas, or oil—releases large amounts of carbon dioxide into the atmosphere. Carbon dioxide is considered a major "greenhouse gas" that contributes to global warming—the gradual increase in the Earth's temperature over time. In addition, coal burning adds sulfur dioxide to the air, which contributes to the formation of acid rain—precipitation that is abnormally acidic. This acid rain can kill forests and leave lakes too acidic to support life. Technology continues to present ways to minimize the pollution that results from extracting and burning fossil fuels. Clean air and climate change policies guide states and industries toward implementing various strategies and technologies for a cleaner coal industry.

Each of the six books in this series—ENVIRONMENTAL ISSUES—introduces the significant points that relate to the specific topic and explains its relationship to other environmental concerns.

Book One: *Air Quality*

Problems of air pollution can be traced back to the time when humans first started to burn coal. *Air Quality* looks at today's challenges in fighting to keep our air clean and safe. The book includes discussions of air pollution sources—car and truck emissions, diesel engines, and many industries. It also discusses their effects on our health and the environment.

The Environmental Protection Agency (EPA) has reported that more than 150 million Americans live in areas that have unhealthy levels of some type of air pollution.[4] Today, more than 20 million Americans, over 6 million of whom are children, suffer from asthma believed to be triggered by pollutants in the air.[5]

In 1970, Congress passed the Clean Air Act, putting in place an ambitious set of regulations to address air pollution concerns. The EPA has identified and set standards for six common air pollutants: ground-level ozone, nitrogen oxides, particulate matter, sulfur dioxide, carbon monoxide, and lead.

The EPA has also been developing the Clean Air Rules of 2004, national standards aimed at improving the country's air quality by specifically addressing the many sources of contaminants. However, many conservation organizations and even some states have concerns over what appears to be an attempt to weaken different sections of the 1990 version of the Clean Air Act. The government's environmental protection efforts take on increasing importance because air pollution degrades land and water, contributes to global warming, and affects the health of plants and animals, including humans.

Book Two: *Climate Change*

Part of science is observing patterns, and scientists have observed a global rise in temperature. *Climate Change* discusses the sources and effects of global warming. Scientists attribute this accelerated change to human activities such as the burning of fossil fuels that emit greenhouse gases (GHG).[6] Since the 1700s, we have been cutting down the trees that help remove carbon dioxide from the atmosphere, and have increased the amount of coal, gas, and oil we burn, all of which add carbon dioxide to the atmosphere. Science tells us that these human activities have caused greenhouse gases—carbon dioxide (CO_2), methane (CH_4), nitrous oxide (N_2O), hydrofluorocarbons (HFCs), perfluorocarbons (PFCs), and sulfur hexafluoride (SF_6)—to accumulate in the atmosphere.[7]

If the warming patterns continue, scientists warn of more negative environmental changes. The effects of climate change, or global warming, can be seen all over the world. Thousands of scientists are predicting rising sea levels, disturbances in patterns of rainfall and regional weather, and changes in ranges and reproductive cycles of plants and animals. Climate change is already having some effects on certain plant and animal species.[8]

Many countries and some American states are already working together and with industries to reduce the emissions of greenhouse gases. Climate change is an issue that clearly fits noted scientist Rene Dubois's advice: "Think globally, act locally."

Book Three: *Conservation*

Conservation considers the issues that affect our world's vast array of living creatures and the land, water, and air they need to survive.

One of the first people in the United States to put the political spotlight on conservation ideas was President Theodore Roosevelt. In the early 1900s, he formulated policies and created programs that addressed his belief that: "The nation behaves well if it treats the natural resources as assets which it must turn over to the next generation increased, and not impaired, in value."[9] In the 1960s, biologist Rachel Carson's book, *Silent Spring*, brought conservation issues into the public eye. People began to see that polluted land, water, and air affected their health. The 1970s brought the creation of the United States Environmental Protection Agency (EPA) and passage of many federal and state rules and regulations to protect the quality of our environment and our health.

Some 80 years after Theodore Roosevelt established the first National Wildlife Refuge in 1903, Harvard biologist

E. O. Wilson brought public awareness of conservation issues to a new level. He warned:

> . . . the worst thing that will probably happen—in fact is already well underway—is not energy depletion, economic collapse, conventional war, or even the expansion of totalitarian governments. As terrible as these catastrophes would be for us, they can be repaired within a few generations. The one process now ongoing that will take million of years to correct is the loss of genetic species diversity by the destruction of natural habitats. This is the folly our descendants are least likely to forgive us.[10]

To heed Wilson's warning means we must strive to protect species-rich habitats, or "hotspots," such as tropical rain forests and coral reefs. It means dealing with conservation concerns like soil erosion and pollution of fresh water and of the oceans. It means protecting sea and land habitats from the over-exploitation of resources. And it means getting people involved on all levels—from national and international government agencies, to private conservation organizations, to the individual person who recycles or volunteers to listen for the sounds of frogs in the spring.

Book Four: *Environmental Policy*
One approach to solving environmental problems is to develop regulations and standards of safety. Just as there are rules for living in a community or for driving on a road, there are environmental regulations and policies that work toward protecting our health and our lands. *Environmental Policy* discusses the regulations and programs that have been crafted to address environmental issues at all levels—global, national, state, and local.

Today, as our resources become increasingly limited, we witness heated debates about how to use our public lands and how to protect the quality of our air and water. Should we allow drilling in the Arctic National Wildlife Refuge? Should we protect more marine areas? Should we more closely regulate the emissions of vehicles, ships, and industries? These policy issues, and many more, continue to make news on a daily basis.

In addition, environmental policy has taken a place on the international front. Hundreds of countries are working together in a variety of ways to address such issues as global warming, air pollution, water pollution and supply, land preservation, and the protection of endangered species. One question the United States continues to debate is whether to sign the 1997 Kyoto Protocol, the international agreement designed to decrease the emissions of greenhouse gases.

Many of the policy tools for protecting our environment are already in place. It remains a question how they will be used— and whether they will be put into action in time to save our natural resources and ourselves.

Book Five: *Water Pollution*

Pollution can affect water everywhere. Pollution in lakes and rivers is easily seen. But water that is out of our plain view can also be polluted with substances such as toxic chemicals, fertilizers, pesticides, oils, and gasoline. *Water Pollution* considers issues of concern to our surface waters, our groundwater, and our oceans.

In the early 1970s, about three-quarters of the water in the United States was considered unsafe for swimming and fishing. When Lake Erie was declared "dead" from pollution and a river feeding it actually caught on fire, people decided that the national government had to take a stronger role in protecting our resources. In 1972, Congress passed the Clean Water Act, a law whose objective "is to restore and maintain the chemical, physical, and biological integrity of the Nation's waters."[11] Today, over 30 years later, many lakes and rivers have been restored to health. Still, an estimated 40% of our waters are still unsafe to swim in or fish.

Less than 1% of the available water on the planet is fresh water. As the world's population grows, our demand for drinking and irrigation water increases. Therefore, the quantity of available water has become a major global issue. As Sandra Postel, a leading authority on international freshwater issues, says, "Water scarcity is now the single biggest threat to global food production." [12] Because there are many competing demands for water, including the needs of habitats, water pollution continues to become an even more serious problem each year.

Book Six: *Wildlife Protection*

For many years, the word *wildlife* meant only the animals that people hunted for food or for sport. It was not until 1986 that the Oxford English Dictionary defined *wildlife* as "the native fauna and flora of a particular region." [13] *Wildlife Protection* looks at overexploitation—for example, overfishing or collecting plants and animals for illegal trade—and habitat loss. Habitat loss can be the result of development, logging, pollution, water diverted for human use, air pollution, and climate change.

Also discussed are various approaches to wildlife protection. Since protection of wildlife is an issue of global concern, it is addressed here on international as well as on national and local levels. Topics include voluntary international organizations such as the International Whaling Commission and the CITES agreements on trade in endangered species. In the United States, the Endangered Species Act provides legal protection for more than 1,200 different plant and animal species. Another approach to wildlife protection includes developing partnerships among conservation organizations, governments, and local people to foster economic incentives to protect wildlife.

CONSERVATION IN THE UNITED STATES

Those who first lived on this land, the Native American peoples, believed in general that land was held in common, not to be individually owned, fenced, or tamed. The white settlers from Europe had very different views of land. Some believed the New World was a Garden of Eden. It was a land of opportunity for them, but it was also a land to be controlled and subdued. Ideas on how to treat the land often followed those of European thinkers like John Locke, who believed that "Land that is left wholly to nature is called, as indeed it is, waste." [14]

The 1800s brought another way of approaching the land. Thinkers such as Ralph Waldo Emerson, John Muir, and Henry David Thoreau celebrated our human connection with nature. By the end of the 1800s, some scientists and policymakers were noticing the damage humans have caused to the land. Leading public officials preached stewardship and wise use of our country's resources. In 1873, Yellowstone National Park was set up. In 1903, the first National Wildlife Refuge was established.

However, most of the government practices until the middle of the 20th century favored unregulated development and use of the land's resources. Forests were clear cut, rivers were dammed, wetlands were filled to create farmland, and factories were allowed to dump their untreated waste into rivers and lakes.

In 1949, a forester and ecologist named Aldo Leopold revived the concept of preserving land for its own sake. But there was now a biological, or scientific, reason for conservation, not just a spiritual one. Leopold declared: "All ethics rest upon a single premise: that the individual is a member of a community of interdependent parts. . . . A thing is right when it tends to preserve the integrity and stability and beauty of the biotic community. It is wrong when it tends otherwise." [15]

The fiery vision of these conservationists helped shape a more far-reaching movement that began in the 1960s. Many credit Rachel Carson's eloquent and accessible writings, such as her 1962 book *Silent Spring*, with bringing environmental issues into people's everyday language. When the Cuyahoga River in Ohio caught fire in 1969 because it was so polluted, it captured the public attention. Conservation was no longer just about protecting land that many people would never even see, it was about protecting human health. The condition of the environment had become personal.

In response to the public outcry about water and air pollution, the 1970s saw the establishment of the EPA. Important legislation to protect the air and water was passed. National standards for a cleaner environment were set and programs were established to help achieve the ambitious goals. Conservation organizations grew from what had started as exclusive white men's hunting clubs to interest groups with a broad membership base. People came together to demand changes that would afford more protection to the environment and to their health.

Since the 1960s, some presidential administrations have sought to strengthen environmental protection and to protect more land and national treasures. For example, in 1980, President Jimmy Carter signed an act that doubled the amount of protected land in Alaska and renamed it the Arctic National Wildlife Refuge. Other administrations, like those of President Ronald Reagan, sought to dismantle many earlier environmental protection initiatives.

The environmental movement, or environmentalism, is not one single, homogeneous cause. The agencies, individuals, and organizations that work toward protecting the environment vary as widely as the habitats and places they seek to protect. There are individuals who begin grass-roots efforts—people like Lois Marie Gibbs, a former resident of the polluted area of Love Canal, New York, who founded the Center for Health, Environment and Justice. There are conservation organizations, like The Nature Conservancy, the World Wildlife Fund (WWF), and Conservation International, that sponsor programs to preserve and protect habitats. There are groups that specialize in monitoring public policy and legislation—for example, the Natural Resources Defense Council and Environmental Defense. In addition, there are organizations like the Audubon Society and the National Wildlife Federation whose focus is on public education about environmental issues. Perhaps from this diversity, just like there exists in a healthy ecosystem, will come the strength and vision environmentalism needs to deal with the continuing issues of the 21st century.

INTERNATIONAL CONSERVATION EFFORTS

In his book *Biodiversity*, E. O. Wilson cautions that biological diversity must be taken seriously as a global resource for three reasons. First, human population growth is accelerating the degrading of the environment, especially in tropical countries. Second, science continues to discover new uses for biological diversity—uses that can benefit human health and protect the environment. And third, much biodiversity is being lost through extinction, much of it in the tropics. As Wilson states, "We must hurry to acquire the knowledge on which a wise policy of conservation and development can be based for centuries to come."[16]

People organize themselves within boundaries and borders. But oceans, rivers, air, and wildlife do not follow such rules. Pollution or overfishing in one part of an ocean can easily degrade the quality of another country's resources. If one country diverts a river, it can destroy another country's wetlands or water resources. When Wilson cautions us that we must hurry to develop a wise conservation policy, he means a policy that will protect resources all over the world.

To accomplish this will require countries to work together on critical global issues: preserving biodiversity, reducing global warming, decreasing air pollution, and protecting the oceans. There are many important international efforts already going on to protect the resources of our planet. Some efforts are regulatory, while others are being pursued by nongovernmental organizations or private conservation groups.

Countries volunteering to cooperate to protect resources is not a new idea. In 1946, a group of countries established the International Whaling Commission (IWC). They recognized that unregulated whaling around the world had led to severe declines in the world's whale populations. In 1986, the IWC declared a moratorium on whaling, which is still in effect, until the populations have recovered.[17] Another example of international cooperation occurred in 1987 when various countries signed the Montreal Protocol to reduce the emissions of ozone-depleting gases. It has been a huge success, and perhaps has served as a model for other international efforts, like the 1997 Kyoto Protocol, to limit emissions of greenhouse gases.

Yet another example of international environmental cooperation is the CITES agreement (the Convention on International Trade in Endangered Species of Wild Fauna and Flora), a legally binding agreement to ensure that the international trade of plants and animals does not threaten the species' survival. CITES went into force in 1975 after 80 countries agreed to the terms. Today, it has grown to include more than 160 countries. This make CITES among the largest conservation agreements in existence.[18]

Another show of international conservation efforts are governments developing economic incentives for local conservation. For example, in 1996, the International Monetary Fund (IMF) and the World Wildlife Fund (WWF) established a program to relieve poor countries of debt. More than 40 countries have benefited by agreeing to direct some of their savings toward environmental programs in the "Debt-for-Nature" swap programs.[19]

It is worth our time to consider the thoughts of two American conservationists and what role we, as individuals, can play in conserving and protecting our world. E. O. Wilson has told us that "Biological Diversity—'biodiversity' in the new parlance—is the key to the maintenance of the world as we know it."[20] Aldo Leopold, the forester who gave Americans the idea of creating a "land ethic," wrote in 1949 that: "Having to squeeze the last drop of utility out of the land has the same desperate finality as having to chop up the furniture to keep warm."[21] All of us have the ability to take part in the struggle to protect our environment and to save our endangered Earth.

ENDNOTES

1   Quammen, David. *Song of the Dodo*. New York: Scribner, 1996, p. 607.

2   Wilson, E. O. *Diversity of Life*. Cambridge, MA: Harvard University Press, 1992, p. 346.

3   Muir, John. *My First Summer in the Sierra*. San Francisco: Sierra Club Books, 1988, p. 110.

4   Press Release. *EPA Newsroom: EPA Issues Designations on Ozone Health Standards.* April 15, 2004. Available online at *http://www.epa.gov/newsroom/.*

5   The Environmental Protection Agency. EPA Newsroom. *May is Allergy Awareness Month.* May 2004. Available online at *http://www.epa.gov/newsroom/allergy_month.htm.*

6   Intergovernmental Panel on Climate Change (IPCC). Third Annual Report, 2001.

7   Turco, Richard P. *Earth Under Siege: From Air Pollution to Global Change.* New York: Oxford University Press, 2002, p. 387.

8   Intergovernmental Panel on Climate Change. *Technical Report V: Climate Change and Biodiversity.* 2002. Full report available online at *http://www.ipcc.ch/pub/tpbiodiv.pdf.*

9   "Roosevelt Quotes." American Museum of Natural History. Available online at *http://www.amnh.org/common/faq/quotes.html.*

10 Wilson, E. O. *Biophilia.* Cambridge, MA: Harvard University Press, 1986, pp. 10–11.

11 Federal Water Pollution Control Act. As amended November 27, 2002. Section 101 (a).

12 Postel, Sandra. *Pillars of Sand.* New York: W. W. Norton & Company, Inc., 1999. p. 6.

13 Hunter, Malcolm L. *Wildlife, Forests, and Forestry: Principles of Managing Forest for Biological Diversity.* Englewood Cliffs, NJ: Prentice-Hall, 1990, p. 4.

14 Dowie, Mark. *Losing Ground: American Environmentalism at the Close of the Twentieth Century.* Cambridge, MA: MIT Press, 1995, p. 113.

15 Leopold, Aldo. *A Sand County Almanac.* New York: Oxford University Press, 1949.

16 Wilson, E. O., ed. *Biodiversity.* Washington, D.C.: National Academies Press, 1988, p. 3.

17 International Whaling Commission Information 2004. Available online at *http://www.iwcoffice.org/commission/iwcmain.htm.*

18 *Discover CITES: What is CITES?* Fact sheet 2004. Available online at *http://www.cites.org/eng/disc/what.shtml.*

19 *Madagascar's Experience with Swapping Debt for the Environment.* World Wildlife Fund Report, 2003. Available online at *http://www.conservationfinance.org/WPC/WPC_documents/Apps_11_Moye_Paddack_v2.pdf.*

20 Wilson, *Diversity of Life*, p. 15.

21 Leopold.

How Will Climate Change Affect the United States?

There is a wide variety of views on climate change. There are people who argue that, as in the case of extinction, climate change will happen over time regardless of what we do. Others contend that instead of warming, the Earth is heading toward a deep freeze. However, most of the scientific community agrees that we have a global problem of climate change because human activities are increasing the rate of change.[1] Scientists use the term *climate change* to mean a change that is caused by human activity, that alters the global atmosphere, and that is observed over comparable time periods as natural climate changes.[2] The danger is that when changes happen quickly, people, other animals, and plants do not have time to adapt. In fact, climate changes could be so extreme that life could not continue.

Since the 1700s, humans have been cutting down forests and increasing our burning of fossil fuels (coal, gas, and oil). Science tells us that this has caused greenhouse gases (carbon dioxide [CO_2], methane [CH_4], nitrous oxide [N_2O], hydrofluorocarbons [HFCs], perfluorocarbons [PFCs], and sulfur hexafluoride [SF_6]) to accumulate in the atmosphere.[3] Based on patterns of observations, scientists predict that increased global warming will lead to rises in sea level, disturbances in patterns of rainfall and regional weather patterns, and changes in plant and animal ranges and reproductive cycles.[4]

The following section is from a chapter in a 2004 report written for the Pew Center for Climate Change. It addresses the potential impact of climate change on the United States. What happens to other parts of the world will depend on their location and their ability to adapt to the changes. The report discusses the potential problems for U.S. agriculture, water resources, coastal communities, human health, land ecosystems, forests, and aquatic (water) ecosystems.

—The Editor

4 Press Release. *EPA Newsroom: EPA Issues Designations on Ozone Health Standards.* April 15, 2004. Available online at *http://www.epa.gov/newsroom/.*

5 The Environmental Protection Agency. EPA Newsroom. *May is Allergy Awareness Month.* May 2004. Available online at *http://www.epa.gov/newsroom/allergy_month.htm.*

6 Intergovernmental Panel on Climate Change (IPCC). Third Annual Report, 2001.

7 Turco, Richard P. *Earth Under Siege: From Air Pollution to Global Change.* New York: Oxford University Press, 2002, p. 387.

8 Intergovernmental Panel on Climate Change. *Technical Report V: Climate Change and Biodiversity.* 2002. Full report available online at *http://www.ipcc.ch/pub/tpbiodiv.pdf.*

9 "Roosevelt Quotes." American Museum of Natural History. Available online at *http://www.amnh.org/common/faq/quotes.html.*

10 Wilson, E. O. *Biophilia.* Cambridge, MA: Harvard University Press, 1986, pp. 10–11.

11 Federal Water Pollution Control Act. As amended November 27, 2002. Section 101 (a).

12 Postel, Sandra. *Pillars of Sand.* New York: W. W. Norton & Company, Inc., 1999. p. 6.

13 Hunter, Malcolm L. *Wildlife, Forests, and Forestry: Principles of Managing Forest for Biological Diversity.* Englewood Cliffs, NJ: Prentice-Hall, 1990, p. 4.

14 Dowie, Mark. *Losing Ground: American Environmentalism at the Close of the Twentieth Century.* Cambridge, MA: MIT Press, 1995, p. 113.

15 Leopold, Aldo. *A Sand County Almanac.* New York: Oxford University Press, 1949.

16 Wilson, E. O., ed. *Biodiversity.* Washington, D.C.: National Academies Press, 1988, p. 3.

17 International Whaling Commission Information 2004. Available online at *http://www.iwcoffice.org/commission/iwcmain.htm.*

18 *Discover CITES: What is CITES?* Fact sheet 2004. Available online at *http://www.cites.org/eng/disc/what.shtml.*

19 *Madagascar's Experience with Swapping Debt for the Environment.* World Wildlife Fund Report, 2003. Available online at *http://www.conservationfinance.org/WPC/WPC_documents/Apps_11_Moye_Paddack_v2.pdf.*

20 Wilson, *Diversity of Life,* p. 15.

21 Leopold.

ISSUES AND CHALLENGES

How Will Climate Change Affect the United States?

There is a wide variety of views on climate change. There are people who argue that, as in the case of extinction, climate change will happen over time regardless of what we do. Others contend that instead of warming, the Earth is heading toward a deep freeze. However, most of the scientific community agrees that we have a global problem of climate change because human activities are increasing the rate of change.[1] Scientists use the term *climate change* to mean a change that is caused by human activity, that alters the global atmosphere, and that is observed over comparable time periods as natural climate changes.[2] The danger is that when changes happen quickly, people, other animals, and plants do not have time to adapt. In fact, climate changes could be so extreme that life could not continue.

Since the 1700s, humans have been cutting down forests and increasing our burning of fossil fuels (coal, gas, and oil). Science tells us that this has caused greenhouse gases (carbon dioxide [CO_2], methane [CH_4], nitrous oxide [N_2O], hydrofluorocarbons [HFCs], perfluorocarbons [PFCs], and sulfur hexafluoride [SF_6]) to accumulate in the atmosphere.[3] Based on patterns of observations, scientists predict that increased global warming will lead to rises in sea level, disturbances in patterns of rainfall and regional weather patterns, and changes in plant and animal ranges and reproductive cycles.[4]

The following section is from a chapter in a 2004 report written for the Pew Center for Climate Change. It addresses the potential impact of climate change on the United States. What happens to other parts of the world will depend on their location and their ability to adapt to the changes. The report discusses the potential problems for U.S. agriculture, water resources, coastal communities, human health, land ecosystems, forests, and aquatic (water) ecosystems.

—The Editor

1. Intergovernmental Panel on Climate Change (IPCC). Third Annual Report, 2001.

2. Hardy, John T. *Climate Change: Causes, Effects, and Solutions*. New York: John Wiley, 2003, p. 11.

3. Turco, Richard P. *Earth Under Siege: From Air Pollution to Global Change*. New York: Oxford University Press, 2002, p. 387.

4. Intergovernmental Panel on Climate Change (IPCC). Technical Report V: Climate Change and Biodiversity.

A Synthesis of Potential Climate Change Impacts on the United States
by Joel B. Smith

NATIONAL IMPACTS BY SECTOR

This section summarizes information on national impacts of climate change. It addresses societal sectors (i.e., sectors heavily managed by society) such as agriculture, water resources, and forestry, and natural sectors such as terrestrial ecosystems and aquatic ecosystems.

In general, societal sectors in the United States are less vulnerable to climate change than natural sectors because the societal sectors have much greater capacity for adaptation.

A. Agriculture
Agriculture is highly sensitive to climate change, but this sector also possesses a substantial capacity to adapt.

Crop yields are likely to be greatly affected by climate change. Changes in temperature, precipitation, and CO_2 concentrations can result in large reductions in growth and water use of some crops and large increases in others. This depends on many factors, including the crops themselves (some are more sensitive to changes in temperature or CO_2 levels than others) and locations (grain crops in more northern locations are more likely to benefit from additional heat, while crops in more southern locations are more likely to have reduced yields).

Agriculture has a significant capacity to adapt to climate change for two reasons. First, agricultural systems can be quickly changed in response to external factors such as change in climate. Second, agriculture is part of a market system. Markets are self-equilibrating because they send signals to producers and consumers about scarcity or abundance of resources. If climate change causes yields to decrease, prices will increase, leading to more production and less demand. If climate change causes yields to increase, prices will fall, leading to less production and more demand.

These two factors combine to substantially moderate the effects of large changes in crop yields. Adams et al. (1999) found that a temperature increase up to 2–3°C (4–5°F) is unlikely to have a significant effect on U.S. agricultural production. Indeed, national production could increase. This is the result of a number of factors, including increased grain yields in the North offsetting reduced grain yields in the South and the ability of agriculture to adapt by introducing new crops and shifting production to areas that become more favorable for crops.

Agricultural output is estimated to peak at a temperature increase of about 2–3°C (4–5°F) and then start declining. Beyond approximately 5°C (9°F), national agricultural production could fall below current levels as a result of the CO_2 effect becoming saturated at higher CO_2 concentrations and increased stress on plants from higher temperatures. Assumptions about the efficiency of adaptation are critical for explaining differences in estimated changes in production among agriculture modeling studies. While agriculture studies have addressed what may happen to production, they have not assessed the costs of making adaptations (e.g., introducing and switching crops, redesigning and relocating infrastructure). They also assume that the adaptations are appropriate, efficient, and effective. In addition, these analyses incorporate optimistic assumptions about CO_2 fertilization. Some of the studies incorporate potential changes in water supplies, while others do not. Also, the studies have not addressed the potential impacts of changes in climate variability or flooding.

While national production may not be significantly affected, there could be large regional changes in agriculture. The Southeast and Southern Great Plains may lose competitive advantage, while more northern areas such as the Midwest and Northern Great Plains could benefit. For example, Adams and McCarl (2001) found that under most of the scenarios they examined, northern agricultural producers had increased output, while southern producers had decreased output. Mendelsohn (2001) found similar results.

B. Water Resources
Although water resources, which are also quite sensitive to climate, have a high potential capacity to adapt, long-lived infrastructure and institutional issues are likely to combine to make adaptation in this sector more challenging than in sectors such as agriculture.

Water resources will be directly affected by climate change through changes in precipitation, evaporation, and snowmelt. While, on average, global precipitation will increase, some areas could see less precipitation, and seasons could be affected differently (e.g., more winter precipitation and less summer precipitation). Temperatures will rise everywhere in the United States, resulting in earlier snowmelt and more evaporation. More frequent and intense floods and droughts are possible. Thus, some regions could benefit from increased supplies, but could face risks from increased flooding. Others could see increased droughts and floods.

The United States has been successful in adapting to quite different hydrologic conditions, ranging from the relatively wet Northeast to the relatively dry Southwest. But it has done so by optimizing infrastructure and institutions such as water laws for current hydrologic conditions. A significant change in climate could make substantial investments in infrastructure or changes in water use patterns and legal arrangements necessary. The time and expense involved in modifying existing infrastructure or building new infrastructure, and the political difficulties involved in changing water laws and other institutional arrangements concerning production and use of water resources could make adaptation of water resources much more challenging than in sectors such as agriculture.

All regions could face risks from changes in water supply and water quality. The Southwest appears to have the greatest sensitivity to changes in water supply because supplies there already are limited. The southern half of the United States has the greatest sensitivity to changes in water quality for a number of reasons, including the already low dissolved oxygen levels, relatively high presence of endangered species, and high sensitivity to low-flow conditions (which is mainly in the Southwest).

C. Coastal Communities
Sea-level rise threatens to inundate and erode many low-lying coastal areas.

The Southeast and mid-Atlantic coasts are the most vulnerable to sea-level rise because they are low-lying, heavily developed in many areas, and at greatest risk from increased hurricane activity. Parts of the Northeast are also vulnerable because they are low-lying and heavily developed. The West Coast is less vulnerable because a large portion of it is made up of rocks or cliffs. However, low-lying areas and bays such as Puget Sound, San Francisco Bay, and much of Southern California are vulnerable to inundation from sea-level rise.

Highly developed coastal areas are likely to be protected from sea-level rise because the costs of protection are lower than the value of property at risk. However, sea-level rise will result in higher infrastructure costs to protect at least some developed coastal areas and in inundation of many unprotected coastal areas. Estimates of the total undiscounted financial costs of adapting to a 0.5 meter [1.6 feet] sea-level rise (roughly the best estimate of eustatic sea-level rise by 2100) range from $20 billion to $138 billion. This includes building coastal defenses to protect high-value areas and abandoning property in low-value areas. The difference in estimates is mainly the result of different assumptions about adaptation. The low estimates assume an economically efficient response (areas are protected when benefits exceed costs and property owners allow properties to depreciate before being inundated), and the high estimates assume that all developed coastal areas will be protected from sea-level rise. Damages will increase with higher magnitudes of sea-level rise. Total (undiscounted) infrastructure costs to protect developed areas from a one meter sea-level rise are estimated to range from $36 billion to $321 billion. These estimates do not consider potential changes in storm or hurricane frequency and intensity; more intense hurricanes would increase risk to life and coastal property.

D. Human Health

Human health is highly sensitive to climate change, but the nation's wealth and its strong public health system will likely prevent any significant human health problems.

Increased flooding, hurricanes, and coastal storms could put more people at risk of injury or death. Heat stress mortality is expected to increase (although increased use of air conditioning can partially offset this), and there could be isolated outbreaks of infectious diseases. Low-income elderly residents of inner cities in the Northeast and Midwest are at greatest risk of heat stress, and increased risk of dengue fever may be greatest along the Rio Grande. However, the complex and nonlinear nature of infectious disease makes it very difficult to predict changes in frequency and location of outbreaks. Should new diseases be introduced, risks may be higher because it may take longer to recognize the presence of the disease and to develop effective preventive measures. Air quality could also deteriorate, particularly if ozone precursors are not further reduced. Cold weather related mortality could decrease, but it is not known whether such a decrease could offset increased mortality from heat stress. Maintaining the public health infrastructure, particularly its capability to monitor for outbreaks of diseases and to intervene when necessary, is a very important factor that may minimize risk of climate change to human health. The costs of maintaining public health under more adverse climate conditions have not been estimated.

Should the rise in temperature be at the upper end of the forecast range in the 21st century or should temperatures continue to rise beyond that, the risks to human health are likely to be greater. The potential for increases in extreme events rises with higher magnitudes of climate change. As average temperatures increase, the potential for more extreme heat waves also rises. Furthermore, at greater magnitudes of climate change there is greater potential for disruption of ecosystems, which could increase the risk of infectious disease outbreaks.

E. Terrestrial Ecosystems

Substantial changes in the distribution of ecosystems in the United States are likely under climate change, with a general shift in species to the north and to higher altitudes, and the potential for the outright elimination of some ecosystems, especially those currently in colder locales.

Climate change will result in a northward or higher altitude shift of climates that species can survive in or are best adapted to. To survive climate change, species will need to migrate because new climates may be beyond their tolerance or may bring competitors, predators, diseases, or other changes that would threaten their survival.

Ecosystems do not migrate as a whole. Instead, individual species migrate—typically at different rates. New ecosystems are likely to involve different assemblages of species. Disturbances such as fire, disease, or drought are likely to play a critical role in changing the location of species. Many species may be unable to migrate in pace with shifting climate zones, may find paths of migration blocked by natural or artificial barriers, or may find themselves outcompeted by invasive species. Even those species that are able to migrate in pace with climate change are likely to encounter other species with which they have not previously interacted. How individual species will respond to novel predator-prey, competitive, symbiotic, or parasitic relationships is largely unknown and difficult to predict. Beyond this, the risk to natural ecosystems from climate change is far more serious because development has put ecosystems under stress through habitat destruction, fragmentation, and pollution. Thus climate change is expected to exacerbate the loss of biodiversity already resulting from development in the United States.

On the other hand, the productivity of terrestrial vegetation in many parts of the United States may increase. Net primary productivity has been estimated to remain the same or increase up to one-third with a doubling of CO_2 concentrations. These estimates are based in part on the assumption that the CO_2 fertilization effect will be fully realized and persistent in natural conditions. However, increased fire frequency and respiration could offset these gains in vegetation productivity. In addition, should the effect of CO_2 on vegetation be less positive, the net effects of climate change on terrestrial ecosystems would be less positive or even negative.

The response of ecosystems is likely to differ from east to west. In the eastern two-thirds of the country, the expected response is a northern shift in location. Parts of the Deep South and upper Midwest could have reduced vegetation productivity. In the western third of the country, mountains may enable species to migrate upslope and provide a north-south corridor. The mountainous topography of the West also means there are many microclimates and barriers to migration—both of which combine to isolate species and inhibit migration. Alpine systems may not survive in the lower 48 states.

As greenhouse gas concentrations continue to increase, there is a greater likelihood of increasingly negative impacts on terrestrial ecosystems. The CO_2 fertilization effect will begin to saturate and the inevitable higher temperatures could cause more stress on species, particularly if precipitation does not increase substantially. Indeed, should temperatures reach the upper end of the projected range in the 21st century, negative effects on most terrestrial ecosystems are likely, particularly if precipitation amounts do not change or decline.

F. Forestry

The location, composition, and abundance of U.S. forests will be altered substantially under climate change, although actual changes will be dependent on future trends in temperature and precipitation and on the ultimate effects of CO_2 fertilization.

Generally, forests will move northward and to higher altitudes, following the same pattern as other terrestrial ecosystems. As with other forms of vegetation, rates of migration are limited and may not keep up with shifts in location of suitable climate zones. Plantation forests may fare much better because they are managed; spacing and rotation periods can be adjusted to account for climate change and species can be substituted after a harvest with more suitable trees for future conditions.

As with agriculture, while there may be substantial differences in how regional forests will be affected by climate change, most economic studies find that the effects on the U.S. forest industry will be relatively small. Gains in one region can offset losses in another. Northern areas are likely to benefit either because yields of species grown there could increase or because more valuable Southern species can be planted in the North. The impacts on Southern and Northwestern forests are mixed: in some scenarios, Southern and Northwestern producers gain, while in others they lose. How the U.S. forest industry fares as a whole depends on a number of factors. Should there be an increase in productivity, consumers will benefit from lower prices, but the industry could be hurt by falling prices unless demand (particularly foreign demand) increases substantially. Should there be decreased yields, the industry would benefit from higher prices, but consumers would be harmed. However, if substitutes for wood become readily available, the industry would be harmed because consumers will start using substitutes, which will keep prices from rising as much as they would otherwise. In addition, the effects of climate change on foreign producers are important. Should forest productivity abroad increase, there could be more imports, benefiting consumers but harming the domestic forest industry.

G. Aquatic Ecosystems
Climate change will greatly alter the character of aquatic ecosystems and could have many adverse effects.

As with terrestrial ecosystems, climate zones and habitats will generally shift northward. A critical issue is the ability of populations to migrate to new locations. The larger and deeper the water bodies, the more feasible it is for species to migrate. Smaller lakes, rivers that do not run north-south or from sufficiently high altitudes, dams, and even some estuaries may block or impede migration to new habitats. Different species have different capabilities to migrate, and new assemblages of species are highly likely. As with terrestrial ecosystems, the greater the impact of development on ecosystems, the more likely it is that the additional impacts of climate change will be negative. Effects will vary, depending on whether freshwater, estuarine, or marine impacts are being considered.

While many warm-water fishes could benefit from lakes and streams becoming more suitable for their survival, cool- and cold-water fish such as trout would find fewer lakes and streams to inhabit. Changes in runoff due to earlier snowmelt and changed precipitation patterns could adversely affect many fishes. Impacts on wetlands will vary depending on types of wetlands, topography, and whether conditions become wetter or drier. Increased summer drought might eliminate or severely contract small wetlands important for migratory waterfowl.

Coastal ecosystems could be harmed if sea levels rise faster than wetlands can accrete sediments or if inland migration of wetlands is blocked. Mangroves may move into southern areas, and the productivity of estuarine ecosystems would be affected by changes in salinity resulting from the combination of sea-level rise and changing runoff. Changes in runoff and winds will also affect circulation in estuaries and aquatic productivity. In addition, 20 to 45 percent of coastal wetlands could be inundated by a 0.5 meter sea-level rise if developed areas are protected, with substantial harm to estuarine ecosystems. With a 1 meter [3.3 feet] sea level rise, 29 to 69 percent of U.S. coastal wetlands would be inundated. The value of these ecosystem impacts is not included in damage estimates.

On the whole, climate change may have more modest effects on open ocean ecosystems than on estuarine or freshwater systems. Limited analysis suggests that productivity of the world's oceans could decline. However, high-latitude areas could benefit. Depending on location, the productivity of oceans is likely to be affected by changes in temperature, the frequency and intensity of events such as the El Niño/Southern Oscillation and the North Atlantic Oscillation, and changes in circulation such as the thermohaline circulation (which includes the Gulf Stream). Many of these changes are currently difficult to predict.

Coral reefs are likely to be harmed by higher temperatures. Many reefs are already being damaged by high sea-surface temperatures caused by El Niño events. In addition, higher atmospheric CO_2 levels reduce calcification in coral reefs, limiting their ability to grow and expand.

How Do Land Preservation Efforts Address the Problems of Global Warming?

The Earth is getting warmer at a much faster rate than can be explained by natural cycles alone. The problem is that if the trend continues, sea levels will rise, rain and weather patterns will change, and more species, unable to adapt, could go extinct.[1] The scientific community generally agrees that human activities are contributing to this warming trend. But there is also much debate over how to prevent global catastrophe. Research is being conducted, countries are working together, and industries are participating in programs to reduce greenhouse gases (GHG).

There is no single answer to the problem of global warming. Using technology to reduce greenhouse gases from industries is one important strategy. But a significant amount of carbon is removed naturally by our forests. Plants use carbon dioxide (CO_2) plus water plus chlorophyll fueled by energy from the sun to make simple sugar (glucose) and then they release oxygen as a by-product. So, as trees grow, they absorb huge amounts of carbon dioxide from the atmosphere. As you might expect, the rain forests, with their fast growth rates, take in great quantities of carbon dioxide.

The following article examines a Nature Conservancy (TNC) program called the Global Climate Change Initiative. The goal is to identify and implement strategies that "will help mitigate greenhouse gas emissions while searching for ways to make natural areas more adaptable to climate change." The program includes projects that will protect more than 1.7 million acres of land in the United States and tropical rain forests. As an example, TNC worked with the Bolivian government, a local foundation, and three energy companies to stop logging on 1.5 million acres in tropical forests.[2]

The Nature Conservancy has been working with communities, businesses, and individuals to protect more than 117 million acres around the world. Founded in 1951, the TNC mission is "to preserve the plants, animals and natural communities that represent the diversity of life on Earth by protecting the lands and waters they need to survive."

—The Editor

1. Intergovernmental Panel on Climate Change (IPCC). Third Annual Report: Climate Change 2001: A Scientific Basis. Available online at *http://www.grida.no/climate/ipcc_tar/wg1/028.htm#e8.*

2. More information on this program is available at The Nature Conservancy Website, online at *http://nature.org/.*

Global Initiative
from The Nature Conservancy

The Earth's climate is changing. The build-up of carbon dioxide and other greenhouse gases in the atmosphere has sent temperatures rising and put the cycles of storms and seasons on an uncertain course. Many species and ecosystems, already at risk from other pressures, will be pushed beyond their natural ability to adapt by the pace and severity of threats now posed by climate change.

Although the predictions are dire, all is not lost. As an organization dedicated to the protection of the diversity of plant and animal life on Earth, The Nature Conservancy is committed to confronting this issue head-on. We are building on our organizational experience in protecting and managing natural areas to help address the causes of climate change and seek solutions that will enable these natural areas to cope.

A WARMING TREND: THE PLANET'S CLIMATE IN TRANSITION
Scientists first wrote of a climatic warming trend in the middle of the 19th century. Since then, they have been forecasting that the planet's climate—in concert with increases in industrialization—is headed for upheaval. They warned that along with global warming would come a larger phenomenon of climate change: ice would melt, seas would rise, storms would intensify and seasons would shift. These warnings, once reserved for the distant future, are quickly becoming the reality of the here and now.

In the 20th century, the world's average surface temperature rose by approximately 6°C (1°F)—a rate greater than in any period over the last 1,000 years. The Intergovernmental Panel on Climate Change (IPCC), an international body of leading scientists, confirmed the cause-and-effect relationship among the actions of people, the increase in temperatures and the alterations in the climate. As the specific cause of the problem, the panel pointed to the combustion of fossil fuels and the destruction of the world's forests as the factors that have led to higher and higher concentrations of heat-trapping greenhouse gases in the Earth's atmosphere. Climate models now predict that these higher concentrations will lead by the end of this century to an increase of 1.4–5.8°C (3–10°F) in the average surface temperature of the Earth, and to a more chaotic climate.

While scientists continue to perfect their predictions, nature already has been showing some signs of the changes that may be in store. Little doubt remains that these changes have the power to degrade habitats, disconnect food chains and drive plants and animals from their current homes.

The Nature Conservancy is measuring the ability of forests to store carbon. Through six "climate action" projects covering 1.7 million acres, we estimate that, over 40 years, the protection and restoration of these largely forested acres will provide a climate benefit equal to keeping 8 million cars off the road for one year.

A PLANET CHANGED: EARTH'S ECOSYSTEMS AT RISK

As temperatures rise and climatic conditions change, many species will not be able to stay in step with their surroundings. In mountainous regions, tree lines are already moving higher. In low-lying arid zones, springs and pools are drying up. Some birds and flying insects have been pushed beyond the limits of their traditional ranges. As oceans warm, coral reefs are bleaching and mangroves and swamplands are being engulfed by rising waters. In addition to these direct effects, a warmer world with a more chaotic climate may do its greatest damage as an accomplice to already established ecological troublemakers. Invasive species, pollution and disease, for example, are threats that might escalate from painful to lethal when given a boost from bad weather. With the pace and severity of climate change, plants and animals must struggle even harder to survive and adapt, and many will not make it on their own.

The Arctic offers one of the most striking examples of the effects envisioned with climate change. With some of the highest rates of warming of any area in the world—winter temperatures have risen 3.9°C (7°F) since the 1950s—Arctic sea ice has shrunk by 7 percent and its thickness has been shaved by more than 3 feet [1 meter] in the past 30 years. This ice is where seals give birth and rear their pups, polar bears thrive and native people fish and hunt. With shifts in the seasons and scarcer, thinner ice, all three inhabitants face an uncertain future.

Climate change poses a serious threat to the full array of ecosystems and their services, including hunting and fishing, as well as all of the others on which people depend. Communities in river valleys, coastal regions and monsoon zones around the world would be affected the most by severe storms and sea-level rise.

Changing weather patterns will likely lead to losses in agricultural production, and to an increased spread of mosquito-borne diseases.

CONFRONTING THE CHALLENGE:
THE GLOBAL CLIMATE CHANGE INITIATIVE

To address these serious issues, The Nature Conservancy is joining with many others who share an interest in ensuring that our planet's ecosystems continue to maintain plant and animal diversity, sustain agricultural and forest production, and mitigate the risk of catastrophic loss from extreme weather events.

The Conservancy, through its Global Climate Change Initiative, is identifying and implementing strategies that will help mitigate greenhouse gas emissions while searching for ways to make natural areas more adaptable to climate change.

The initiative focuses on the nexus between environmental conservation and climate change through three principal approaches:

1) **Protecting and Restoring Native Forests, Grasslands and Other Ecosystems**—An effective approach to the problem of climate change must include reductions in industrial emissions and increases in the conservation and restoration of native forests and grasslands. In light of its 50 years of experience protecting land, the Conservancy is focused on the latter. Forests store carbon in their leaves, wood, roots and soils. As trees grow they remove the greenhouse gas carbon dioxide from the air, storing it as additional carbon in their tissues as part of the process of photosynthesis. When they are destroyed, so ends their ability to store carbon dioxide, and the gas is released back into the atmosphere. At present, deforestation accounts for one-quarter of annual carbon dioxide emissions, while the protection and restoration of forests may be able to offset up to 20 percent of carbon dioxide emissions over the next 50 years. The Conservancy is putting this idea into practice by implementing forest protection and restoration projects called "climate action" projects, including 6 in the United States, Belize, Bolivia and Brazil. These projects safeguard more than 1.7 million acres while reducing the build-up of greenhouse gases in the atmosphere.

2) **Adapting to the Change**—The Conservancy is examining the effect of climate change on our ability to follow the four steps of Conservation by Design, our plan for conservation success.

   • Along the Arctic Coast of Alaska, we are studying the susceptibility of caribou and other target species to climatic changes, identifying areas that may offer them refuge and using that information when setting priorities for the surrounding ecoregion.

- In Noel Kempff Mercado National Park in Bolivia, we are developing strategies to cope with climate change, The Conservancy is adapting our conservation approaches to take into account the ways climate change may affect the habitat of species such as the golden snub-nosed monkey (*Rhinopithecus bieti*) in the Yunnan Province of China.

THE GLOBAL CLIMATE CHANGE INITIATIVE AT WORK
Noel Kempff Mercado National Park, Bolivia

When 1.5 million acres of tropical forest adjacent to this national park in northeastern Bolivia were threatened with timber harvesting and deforestation, The Nature Conservancy turned the threat into an opportunity to offset carbon dioxide emissions.

By facilitating a unique partnership among the Bolivian government, our local partner Fundación Amigos de la Naturaleza and three energy companies, we helped terminate the logging rights and the land was incorporated into the national park.

This "climate action" project doubled the protected habitat for wide-ranging species such as jaguar and averted soil runoff into the region's many rivers while reducing net global carbon dioxide emissions.

Although the park is helping combat climate change, it is not immune from the effects of climate change. With this in mind, the Conservancy is using a variety of tools—from satellite imagery to on-the-ground studies—to monitor changes in the park's tropical forests and savannas, home to jaguars, macaws, monkeys, giant otters, river dolphins, rare plants and many species yet to be described. The hypothesis is that today's monitoring will help park managers anticipate tomorrow's changes and ensure the protection of this last great place.

In Noel Kempff Mercado National Park in Bolivia, the Conservancy has worked to reduce greenhouse gas emissions by preventing deforestation, and is planning for the effects of climate by including it as a factor in park management and improving the management of the park by using satellite technology to assess and anticipate the effects of climate change on tropical forests and savannas.

- In the Albermarle Sound of North Carolina and Virginia in the United States, we are taking action now to preserve the composition of existing natural areas, and also, in anticipation of predicted sea level rise, protecting places that will enable species and their surrounding systems to migrate inland and upstream.

- In the Yunnan Province of China, we are paying particular attention to the ways in which climate change may impede our ability to measure the success of programs designed to protect and restore native ecosystems.

The lessons we learn from these four projects and others like them will inform our efforts as well as the work of other conservationists around the world.

3) **Creating a Policy Framework**—To be truly effective, policies developed to address climate change must account for the role played by forests and other natural systems. The Conservancy is therefore working to make the conservation and restoration of natural areas a more widely known and broadly applied method for reducing the amount of carbon dioxide in the atmosphere while protecting critical plant and animal habitats.

What Is One Conservation Organization Doing to Slow Climate Change?

It is true that air pollution happens in nature. Forest fires, volcanoes, and dust storms put particulates and gases into the air. Plants, in their process of respiration, give off volatile organic compounds. But it comes down to numbers. As the following article from Conservation International (CI) explains, humans put the same amount of carbon dioxide into the air every two days as the eruption of volcano Mount St. Helens released on May 18, 1980. Humans burning coal, gas, and oil add most of this carbon dioxide, a greenhouse gas (GHG) that is contributing to global warming. The solution seems obvious: We must cut the harmful emissions. Toward this end, in 1992, about 182 countries agreed to reduce their GHG emissions.

Somewhat like an algebra equation you are trying to balance, there are activities that put carbon dioxide into the atmosphere, and processes that remove it from the system. The latter are called sinks. The oceans remove about 30% of the carbon dioxide humans emit into the atmosphere. But, as a 2004 U.S. government report states, "the future behavior of this important carbon sink is quite uncertain because of the potential climate change impacts on ocean circulation, biogeochemical cycling, and ecosystem dynamics."[1]

Another important carbon sink is our forests. According to the following Conservation International information, about one-fourth of the necessary carbon reductions could be achieved if we stopped cutting down the rain forests. Despite this information, forests are being cut down at an alarming rate. Fewer trees to use carbon dioxide in photosynthesis (the process of plant respiration) means more carbon remains in the atmosphere.

As a result, some environmental groups, such as Conservation International, are working with industries to help them reduce carbon dioxide emissions. CI also helps design investment plans for the money saved in land-based projects that use trees as carbon dioxide sinks. As the following article states, the goal is to develop sound financial investment programs that help both people and protect habitats and biodiversity.

As science continues to give us more information about how our planet works, it becomes clear that when one global problem, such as carbon dioxide emissions, is addressed, other issues, like biological diversity, benefit.

—The Editor

1. Doney, S. C., et al. *Ocean Carbon and Climate Change (OCCC): An Implementation Strategy for U.S. Ocean Carbon Cycle Science.* Boulder, CO: UCAR, 2004, p. 3.

2. Conservation International Home Page Website. Available online at *http://www.conservation.org/xp/CIWEB/home*.

Weathering Climate Change
from Conservation International

Global warming has begun, the battle to fight it has been joined, and the enemy—no surprise—is us. How would people react if every other day they awoke to media coverage on another devastating volcanic eruption the size of Mount St. Helens? In terms of equivalent units of carbon dioxide (CO_2), humanity is spewing this amount of greenhouse gases (GHG) into the atmosphere every two days.

The scientific consensus is that this thickening layer of heat-trapping gases is affecting the global climate and regional weather patterns. Those changes, in turn, are affecting biodiversity. Scientists have already observed substantial and systematic effects on a wide range of species and ecosystems, and this situation is expected to significantly worsen in the decades ahead. In the past 30 years, numerous shifts in the distribution and abundance of species have resulted from climate change. Adverse effects range from the disappearance of toads in Costa Rica's cloud forests, to the death of coral reefs throughout the planet's tropical marine environments.

SCRUBBING THE SKIES: HOW TO CHILL A GLOBAL GREENHOUSE

Combustion of coal, oil, and natural gas accounts for most global CO_2 emissions. Climate discussions and international policy usually focus on reducing these emissions by shifting to climate-friendly energy options. Less well known is the fact that trees absorb CO_2 and release it when cut down. The 34 million acres of tropical forests destroyed each year—a combined area the size of Ohio—release 20 to 25 percent of total global CO_2 emissions.

The international community first recognized the threat of climate instability at the 1992 United Nations Earth Summit in Rio de Janeiro. Some 182 nations, including the United States, agreed to voluntarily reduce GHGs emitted by industrial countries to their 1990 levels by 2012. More importantly, the signatories agreed to stabilize GHG atmospheric concentrations to prevent dangerous anthropogenic, or human, interference with the climate system.

Since then, scientists estimate that stabilizing atmospheric concentrations of CO_2 at a safe level will require reducing CO_2 emissions and other heat-trapping gases to 80 percent of the 1990 global levels. According to some scenarios, this means reducing or preventing the release of 1.2 trillion tons of CO_2 by 2050. Energy-efficiency improvements throughout the global economy could prevent one-third or more of these emissions while also cutting energy costs.

Roughly one-fourth of the necessary reductions, about 370 billion tons, could be achieved by preventing tropical deforestation, regenerating fragmented landscapes into connected conservation corridors, restoring several billion acres of degraded lands, and improving land productivity worldwide through the adoption of best practices in agriculture and forestry. Taking advantage of selective energy supply options such as solar or wind power that have no adverse effects on climate and biodiversity could help reduce the remaining energy-related emissions.

TAKING ACTION: CI ADJUSTS STRATEGY
TO ACCOMMODATE A WARMER WORLD

Strengthening existing conservation efforts not only helps to mitigate the threat of climate destabilization, but also offers greater resilience against weather disasters that threaten both people and habitat. CI already has a solid record in creating protected areas and conservation corridors. In the past three years, for instance, our Centers for Biodiversity Conservation have helped establish or improve the management of roughly 55 million acres of parks and protected areas in the biodiversity hotspots and high-biodiversity wilderness areas.

However, species' ranges will adjust in response to climate change, and this will influence our ability to conserve them in existing parks. Shifting of range boundaries because of warming has already taken place during the past 30 to 70 years. Expanding our conservation strategies to meet that challenge is now central to CI's mission, and we believe this will significantly reduce the extinction risk associated with global warming.

Researchers from the Center for Applied Biodiversity Science (CABS) at CI, in partnership with The Nature Conservancy, are leading efforts to anticipate the effects of climate change on biodiversity, and to integrate that knowledge into our conservation planning. CABS research has concentrated on South Africa's Cape Floristic Region, a biodiversity hotspot and one of the world's six plant kingdoms. A pioneering multi-species modeling study conducted by CABS and South African researchers reveals how climate change might influence conservation efforts in protected areas. It shows which species will be most affected, whether their numbers will shrink or grow, and how the changes will unfold over time. Such information will allow conservationists to add new parks and corridors that are more resilient to wild weather swings.

TAPPING EMERGING CARBON MARKETS: A HUGE PROFIT POTENTIAL

No discussion of climate change is complete without determining ways to mitigate GHG emissions from the burning of fossil fuels. The Center for Environmental Leadership in Business at CI is working with industry to reduce companies' CO_2 emissions and enhance their competitive advantage. The strategy employs energy-efficiency improvements like solar or wind power, and the money saved can be invested in multi-benefit land-based CO_2 offsets.

These projects absorb CO_2 from the atmosphere and store it in trees, other vegetation, and soil. Well-designed and well-executed offsets can help counter climate change and sustain protected wilderness habitats while fostering alternative, sustainable livelihoods among impoverished local communities. Conversely, poorly designed monoculture projects that promote large areas planted with a single tree species, can harm biodiversity, disrupt watersheds, and have little impact on sustainable development.

Launched in 2002, CI's offset strategy allows companies to invest in CO_2 mitigation while both contributing to conservation and supporting local economies. This approach offers a number of important benefits, including biodiversity protection and improved human welfare. Many cash-strapped developing countries have forests that are carbon- and biodiversity-rich habitats. By not logging or razing their forests, developing countries could market these carbon services to companies and to industrial nations seeking cost-effective ways to offset their high emission levels, either voluntarily or for regulatory compliance requirements.

In addition to CI's offset program, the Climate, Community, and Biodiversity Alliance initiative has been launched to develop standards for benefits of land-based offset projects. Several companies with strong environmental leadership records have joined the alliance. They include BP [British Petroleum], the world's second largest energy company; Intel, the world's largest semiconductor manufacturer; and SC Johnson, the Fortune 500 household products company. Creating sound CO_2 mitigation investments that help people and biodiversity could potentially bring in billions of dollars for biodiversity conservation and struggling economies. Will this strategy reverse the global warming trend? By itself, no—much more needs to be done in industrial countries to reduce CO_2 emissions. However, it will help stabilize GHGs, while countering the massive deforestation now taking place in the hotspots and wilderness areas. Combined with research to anticipate habitat changes brought on by climate change, such investments will help life on Earth weather the coming century.

Current & Projected Effects of Climate Change
from Conservation International

CURRENT EFFECTS
Climate change is already affecting global biodiversity, and scientists expect the impacts to increase.

- Coral reefs are extremely vulnerable to global warming. They contain microorganisms that are expelled when water temperatures rise, causing the corals to lose color or bleach. More than 16 percent of the planet's coral reefs have been severely damaged in recent years because of unusually high water temperatures, and many have died.

- Global warming has caused the loss of some 12 billion cubic meters [15.7 billion cubic yards] of snow from Peruvian Andes glaciers, where 70 percent of the world's tropical region ice fields are located. The fate of these glaciers will affect water supplies and ecological habitats across the vast Amazon region.

- The disappearance of 20 species of frogs and toads, including the endemic golden toad from the highland cloud forests in Costa Rica, has been linked to a 30-year warming trend and a severe reduction in dry-season mists. Meanwhile, species from lower elevations are invading the forests.

- A glacial retreat in high-biodiversity regions such as eastern Africa is imperiling watersheds. During the past century, glaciers on Mount Kilimanjaro have lost 73 percent of their mass. Glacial melt water nourishes many alpine ecosystems, and certain desert streamside habitats would completely collapse without it in the dry season.

PROJECTED EFFECTS

A variety of habitats from alpine to marine will be modified by global warming.

- Reduced habitat will place some species at greater risk of extinction. They include the Bengal tiger in the Sundarbans Delta region of India and Bangladesh, the mountain gorilla in Africa, and the resplendent quetzal in Central and South America.

- Sea-level rise, caused by increasing ocean temperatures, is expected to result in the disappearance of many low-lying islands along with the extinction of their endemic species. Two South Pacific islands in the state of Kiribati have already disappeared beneath the waves.

- The warming of tropical waters may contribute to the outbreak of animal (species-based) epidemics and may increase the spread of marine disease agents and parasites. One epidemic killed thousands of striped dolphins in the Mediterranean in the early 1990s.

- Climate change is expected to disrupt many ecosystems by making them susceptible to invasion by non-native species. The anticipated spread of red fire ants throughout the southeastern United States, for example, could devastate native flora and fauna.

Do Most Scientists Agree That Climate Change Is a Global Problem?

Air, like the oceans, does not recognize national boundaries or follow any laws besides the laws of nature. Neither does pollution in the air. And it is the human activities that release different gases and chemicals into the air that are contributing to climate change. The World Meteorological Organization (WMO) and the United Nations Environmental Programme (UNEP) have recognized that climate change is an issue that must be addressed on a global level.

In 1988, the WMO and UNEP established an organization called the Intergovernmental Panel on Climate Change (IPCC) to address growing concerns about the changing climate. Today, the IPCC is regarded as "the international authority for conducting assessments of the current state of knowledge about climate change."[1] In fact, the IPCC Second Assessment Report, issued in 1995, provided key information used in the adoption of the 1997 Kyoto Protocol, an agreement that commits the countries that sign it to establish targets for reducing greenhouse gases.[2]

The IPCC Third Assessment Report (TAR) was released in 2001 and represents the collective work of more than 2,000 experts from a range of disciplines and countries.[3] The findings of the TAR point to the problems of a world that is getting warmer. Dr. Robert Watson, the chair of the IPCC in 2001, stated in the report, "The overwhelming majority of scientific experts, whilst recognizing that scientific uncertainties exist, nonetheless believe that human-induced climate change is already occurring and that future change is inevitable."[4]

The following sections are from one part of the TAR 2001 Synthesis Report. The language is formal and sometimes challenging, but it is worth the time to consider the type of scientific thinking that shapes world environmental policy. The first question considered is the effect of greenhouse gases over the next century if nothing is done. The report addresses the negative effects on humans and the environment. The second question looks at the connection between climate change caused by humans and other

27

environmental issues—for example, air pollution, acid rain, and loss of biological diversity (plant and animal species). Experts agree that there is serious potential for damage to our planet.

—The Editor

1. IPCC Press Advisory. Geneva/Paris. February 14, 2004. Available online at *http://www.ipcc-nggip.iges.or.jp/index.html*.

2. IPCC. Introduction to the IPCC. Available online at *http://www.ipcc-nggip. iges.or.jp/index.html*.

3. IPCC Press Advisory. Milan. December 8, 2003. Available online at *http://www.ipcc-nggip.iges.or.jp/index.html*.

4. IPCC Third Annual Report: Climate Change 2001. Intergovernmental Panel on Climate Change.

IPCC Third Assessment Report: Climate Change 2001: Synthesis Report: Summary for Policymakers
from the Intergovernmental Panel on Climate Change (IPCC)

What is known about the regional and global climatic, environmental, and socio-economic consequences in the next 25, 50, and 100 years associated with a range of greenhouse gas emissions arising from scenarios used in the TAR (projections which involve no climate policy intervention)? To the extent possible evaluate the:

• Projected changes in atmospheric concentrations, climate, and sea level

• Impacts and economic costs and benefits of changes in climate and atmospheric composition on human health, diversity and productivity of ecological systems, and socio-economic sectors (particularly agriculture and water)

• The range of options for adaptation, including the costs, benefits, and challenges

- Development, sustainability, and equity issues associated with impacts and adaptation at a regional and global level.

- Carbon dioxide concentrations, globally averaged surface temperature, and sea level are projected to increase under all IPCC emissions scenarios during the 21st century.

For the six illustrative SRES [Special Report on Emissions Scenarios] emissions scenarios, the projected concentration of CO_2 in the year 2100 ranges from 540 to 970 ppm [parts per million], compared to about 280 ppm in the pre-industrial era and about 368 ppm in the year 2000. The different socio-economic assumptions (demographic, social, economic, and technological) result in the different levels of future greenhouse gases and aerosols. Further uncertainties, especially regarding the persistence of the present removal processes (carbon sinks) and the magnitude of the climate feedback on the terrestrial biosphere, cause a variation of about -10 to +30% in the year 2100 concentration, around each scenario. Therefore, the total range is 490 to 1,250 ppm (75 to 350% above the year 1750 (preindustrial) concentration). Concentrations of the primary non-CO_2 greenhouse gases by year 2100 are projected to vary considerably across the six illustrative SRES scenarios.

Projections using the SRES emissions scenarios in a range of climate models result in an increase in globally averaged surface temperature of 1.4 to 5.8°C [34.5 to 42.4°F] over the period 1990 to 2100. This is about two to ten times larger than the central value of observed warming over the 20th century and the projected rate of warming is very likely to be without precedent during at least the last 10,000 years, based on paleoclimate data. Temperature increases are projected to be greater than those in the Second Assessment Report (SAR), which were about 1.0 to 3.5°C [33.8 to 38.3°F] based on six IS92 ["IS" stands for "integrated science"] scenarios. The higher projected temperatures and the wider range are due primarily to lower projected sulfur dioxide (SO_2) emissions in the SRES scenarios relative to the IS92 scenarios. For the periods 1990 to 2025 and 1990 to 2050, the projected increases are 0.4 to 1.1°C [32.7 to 34°F] and 0.8 to 2.6°C [33.4 to 36.7°F] , respectively. Projections of changes in climate variability, extreme events, and abrupt/non-linear changes are covered [below].

SUMMARY FOR POLICYMAKERS

By the year 2100, the range in the surface temperature response across different climate models for the same emissions scenario is comparable to the range across different SRES emissions scenarios for a single climate model. . . . Nearly all land areas will very likely warm more than these global averages, particularly those at northern high latitudes in winter.

Globally averaged annual precipitation is projected to increase during the 21st century, though at regional scales both increases and decreases are projected of typically 5 to 20%. It is likely that precipitation will increase over high-latitude regions in both summer and winter. Increases are also projected over northern mid-latitudes, tropical Africa, and Antarctica in winter, and in southern and eastern Asia in summer. Australia, Central America, and southern Africa show consistent decreases in winter rainfall. Larger year-to-year variations in precipitation are very likely over most areas where an increase in mean precipitation is projected.

Glaciers are projected to continue their widespread retreat during the 21st century. Northern Hemisphere snow cover, permafrost, and sea-ice extent are projected to decrease further.

The Antarctic ice sheet is likely to gain mass, while the Greenland ice sheet is likely to lose mass (see below).

Global mean sea level is projected to rise by 0.09 to 0.88 m [0.3 to 2.89 feet] between the years 1990 and 2100, for the full range of SRES scenarios, but with significant regional variations. This rise is due primarily to thermal expansion of the oceans and melting of glaciers and ice caps. For the periods 1990 to 2025 and 1990 to 2050, the projected rises are 0.03 to 0.14 m [0.1 to 0.46 feet] and 0.05 to 0.32 m [0.16 to 1.05 feet], respectively.

Projected climate change will have beneficial and adverse effects on both environmental and socio-economic systems, but the larger the changes and rate of change in climate, the more the adverse effects predominate.

The severity of the adverse impacts will be larger for greater cumulative emissions of greenhouse gases and associated changes in climate (medium confidence). While beneficial effects can be identified for some regions and sectors for small amounts of climate change, these are expected to diminish as the magnitude of climate change increases. In contrast many identified adverse effects are expected to increase in both extent and severity with the degree of climate change. When considered by region, adverse effects are projected to predominate for much of the world, particularly in the tropics and subtropics.

Overall, climate change is projected to increase threats to human health, particularly in lower income populations, predominantly within tropical/subtropical countries.

Climate change can affect human health directly (e.g., reduced cold stress in temperate countries but increased heat stress, loss of life in floods and storms) and indirectly through changes in the ranges of disease vectors (e.g., mosquitoes), water-borne pathogens, water quality, air quality, and food availability and quality (medium to high confidence). The actual health impacts will be strongly influenced by local environmental conditions and socio-economic circumstances, and by the range of social, institutional, technological, and behavioral adaptations taken to reduce the full range of threats to health.

Ecological productivity and biodiversity will be altered by climate change and sealevel rise, with an increased risk of extinction of some vulnerable species (high to medium confidence). Significant disruptions of ecosystems from disturbances such as fire, drought, pest infestation, invasion of species, storms, and coral bleaching events are expected to increase. The stresses caused by climate change, when added to other stresses on ecological systems, threaten substantial damage to or complete loss of some unique systems and extinction of some endangered species. The effect of increasing CO_2

concentrations will increase net primary productivity of plants, but climate changes, and the changes in disturbance regimes associated with them, may lead to either increased or decreased net ecosystem productivity (medium confidence). Some global models project that the net uptake of carbon by terrestrial ecosystems will increase during the first half of the 21st century but then level off or decline.

Models of cereal crops indicate that in some temperate areas potential yields increase with small increases in temperature but decrease with larger temperature changes (medium to low confidence). In most tropical and subtropical regions, potential yields are projected to decrease for most projected increases in temperature (medium confidence). Where there is also a large decrease in rainfall in subtropical and tropical dryland/rainfed systems, crop yields would be even more adversely affected. These estimates include some adaptive responses by farmers and the beneficial effects of CO_2 fertilization, but not the impact of projected increases in pest infestations and changes in climate extremes. The ability of livestock producers to adapt their herds to the physiological stresses associated with climate change is poorly known. Warming of a few °C or more is projected to increase food prices globally, and may increase the risk of hunger in vulnerable populations.

Climate change will exacerbate water shortages in many water-scarce areas of the world. Demand for water is generally increasing due to population growth and economic development, but is falling in some countries because of increased efficiency of use. Climate change is projected to substantially reduce available water (as reflected by projected runoff) in many of the water-scarce areas of the world, but to increase it in some other areas (medium confidence). Freshwater quality generally would be degraded by higher water temperatures (high confidence), but this may be offset in some regions by increased flows.

The aggregated market sector effects, measured as changes in gross domestic product (GDP), are estimated to be negative for many developing countries for all magnitudes of global mean temperature increases studied (low confidence), and are estimated to be mixed for developed countries for up to a few °C warming (low confidence) and negative for warming beyond a few degrees (medium to low confidence). The estimates generally exclude the effects of changes in climate variability and extremes, do not account for the effects of different rates of climate change, only partially account for impacts on goods and services that are not traded in markets, and treat gains for some as canceling out losses for others.

Populations that inhabit small islands and/or low-lying coastal areas are at particular risk of severe social and economic effects from sea-level rise and storm surges.

Many human settlements will face increased risk of coastal flooding and erosion, and tens of millions of people living in deltas, in low-lying coastal areas, and on small islands will face risk of displacement. Resources critical to island and coastal populations such as beaches, freshwater, fisheries, coral reefs and atolls, and wildlife habitat would also be at risk.

The impacts of climate change will fall disproportionately upon developing countries and the poor persons within all countries, and thereby exacerbate inequities in health status and access to adequate food, clean water, and other resources.

Populations in developing countries are generally exposed to relatively high risks of adverse impacts from climate change. In addition, poverty and other factors create conditions of low adaptive capacity in most developing countries.

Adaptation has the potential to reduce adverse effects of climate change and can often produce immediate ancillary benefits, but will not prevent all damages.

SUMMARY FOR POLICYMAKERS

Numerous possible adaptation options for responding to climate change have been identified that can reduce adverse and enhance beneficial impacts of climate change, but will incur costs. Quantitative evaluation of their benefits and costs and how they vary across regions and entities is incomplete.

Changes in runoff are calculated with a hydrologic model using as inputs climate projections from two versions of the Hadley Centre atmosphere-ocean general circulation model (AOGCM) for a scenario of 1% per annum increase in effective CO_2 concentration in the atmosphere: (a) HadCM2 ensemble mean and (b) HadCM3. Projected increases in runoff in high latitudes and southeast Asia and decreases in central Asia, the area around the Mediterranean, southern Africa, and Australia are broadly consistent across the Hadley Centre experiments, and with the precipitation projections of other AOGCM experiments. For other areas of the world, changes in precipitation and runoff are scenario- and model-dependent.

Greater and more rapid climate change would pose greater challenges for adaptation and greater risks of damages than would lesser and slower change. Natural and human systems have evolved capabilities to cope with a range of climate variability within which the risks of damage are relatively low and ability to recover is high. However, changes in climate that result in increased frequency of events that fall outside the historic range with which systems have coped increase the risk of severe damages and incomplete recovery or collapse of the system.

QUESTION 4

What is known about the influence of the increasing atmospheric concentrations of greenhouse gases and aerosols, and the projected human-induced change in climate regionally and globally on: The frequency and magnitude of climate fluctuations, including daily, seasonal, inter-annual, and decadal variability, such as the El Niño Southern Oscillation cycles and others? The duration, location, frequency, and intensity of extreme events such as heat waves, droughts, floods, heavy precipitation, avalanches, storms, tornadoes, and tropical cyclones? The risk of abrupt/non-linear changes in, among others, the sources and sinks of greenhouse gases, ocean circulation, and the extent of polar ice and permafrost? If so, can the risk be quantified? The risk of abrupt or non-linear changes in ecological systems? An increase in climate variability and some extreme events is projected. Models project that increasing atmospheric concentrations of greenhouse gases will result in changes in daily, seasonal, inter-annual, and decadal variability. There is projected to be a decrease in diurnal temperature range in many areas, decrease of daily variability of surface air temperature in winter, and increased daily variability in summer in the Northern Hemisphere land areas. Many models project more El Niño–like mean conditions in the tropical Pacific. There is no clear agreement concerning changes in frequency or structure of naturally occurring atmosphere-ocean circulation patterns such as that of the North Atlantic Oscillation (NAO).

Models project that increasing atmospheric concentrations of greenhouse gases result in changes in frequency, intensity, and duration of extreme events, such as more hot days, heat waves, heavy precipitation events, and fewer cold days. Many of these projected changes would lead to increased risks of floods and droughts in many regions, and predominantly adverse impacts on ecological systems, socio-economic sectors, and human health. High resolution modeling studies suggest that peak wind and precipitation intensity of tropical cyclones are likely to increase over some areas. There is insufficient information on how very small-scale extreme weather phenomena (e.g., thunderstorms, tornadoes, hail, hailstorms, and lightning) may change.

What Effect Could Global Warming Have on Air Pollution?

Medical experts are saying that as climate changes bring on global warming, the ozone concentrations in the air are increasing and causing health problems. Higher temperatures cause more smog and more respiratory illnesses.

By definition, ozone is made up of three linked atoms of oxygen, O_3. It is true that the ozone layer protects us from the harmful ultraviolet (UV) rays of the sun. But that is the ozone in the stratosphere, about 10 to 20 miles [16 to 32 km] above the Earth's surface. Ozone closer to home, in the troposphere, is a health hazard when we breathe it. Heat and sunlight increase the reactions that form ozone from the air pollutants nitrogen oxides (NOx) and volatile organic compounds (VOCs). So, as the planet warms from greenhouse gases, the conditions for forming ground level ozone increase. In addition to the warming from greenhouse gases, any holes in the protective stratospheric ozone layer allow even more UV light from the sun to reach the planet, which also powers the reaction to make more ground-level ozone.

—The Editor

Heat Advisory: How Global Warming Causes More Than Bad Air Days
from the Natural Resources Defense Council

CLIMATE CHANGE AND AIR POLLUTION

More than 100 million Americans currently live in counties that do not comply with health-based air quality standards for ground-level ozone, commonly known as smog. Although many types of air pollution have been reduced in the United States since the 1970 Clean Air Act, improvements in ozone pollution levels have slowed since the 1990s. As a result, smog has become a persistent environmental health problem that

36

can aggravate allergies and respiratory illness, particularly in children and the elderly.

What's more, ozone levels are more sensitive to temperature and weather than other air pollutants, suggesting that global warming could worsen the health risks from ozone. Rising temperatures across the eastern United States will increase smog, and this in turn could mean more hospital admissions from respiratory illnesses such as asthma.

In short, the increase in air pollution brought on by global warming may choke the Clean Air Act's goal to provide all Americans with clean, healthy air to breathe.

OZONE AS AN AIR POLLUTANT

Air pollution is the contamination of air by gases and particulates. Air pollutants result from both natural and human activities. Human causes of air pollution include the burning of fossil fuels for transportation and power generation, industrial processes, and garbage incineration. Examples of air pollutants emitted by both natural processes and human activities include aerosols, nitrogen oxides, nonmethane volatile organic compounds, carbon monoxide, carbon dioxide, and methane.

The U.S. Environmental Protection Agency (EPA) has identified six types of air pollutants as "criteria" pollutants that are harmful to the public health and the environment:

- Carbon monoxide (CO)

- Lead (Pb)

- Nitrogen dioxide (NO_2)

- Ozone (O_3)

- Particulate matter (PM)

- Sulfur dioxide (SO_2).

Under the Clean Air Act, National Ambient Air Quality Standards (NAAQS) are established for these criteria air pollutants at levels designed to protect public health. There are currently no federal regulations limiting emissions or concentrations of carbon dioxide, the heat-trapping pollutant primarily responsible for global warming.

What Is Ozone?

Ozone is a colorless gas that is present in both the upper atmosphere (stratosphere) and the lower atmosphere (troposphere). Stratospheric ozone forms a protective layer around the earth that absorbs harmful ultraviolet radiation (UV-B). Nearer to the ground, in the troposphere, a significant amount of ozone is formed when its chemical precursors—nitrogen oxides (NOx) and volatile organic compounds (VOCs)—interact in the atmosphere in the presence of heat and sunlight.

Global emissions of nitrogen oxides are associated mainly with human activities: fossil fuel use, biomass burning, and aircraft. Natural sources for NOx include microbial activity in soils, lightning, and transport from the stratosphere. Both ozone and particulate matter formed in the atmosphere (secondary aerosols) may share common precursor species such as NOx. . . . As stratospheric ozone is depleted, more ultraviolet radiation reaches the earth's surface, where it can affect tropospheric ozone, especially in polluted regions that have abundant ozone precursor gases. This is because heat and ultraviolet light are required in the chemical reaction between NOx and VOCs to form ozone.

Patterns of ozone concentrations on the earth's surface are affected by a number of factors, including emission densities, seasonality, changing weather conditions, and atmospheric transport. Ozone, along with its precursors, can be carried hundreds of miles from the original pollution source by wind and weather (called *atmospheric transport*). Temperature, cloudiness, and atmospheric transport modify the impact of UV-B. And changes in the chemical composition of the atmosphere, including aerosols, will have an impact on ozone concentrations.

Ozone Levels in the United States

According to the EPA, ozone levels have decreased over the past 10 to 25 years as a result of emission controls mandated by the Clean Air Act. However, the downward trend is slowing. What's more, an EPA ozone report says:

> Nationally, 2003 was a good year for ozone air quality. Much of the good news can be attributed in part to favorable weather conditions across many parts of the nation. . . . For the eastern half of the country, the height of ozone season (June through August) was cooler and wetter than normal; therefore, it was less conducive to the formation of ozone than in past years.

In fact, the EPA says that the number of summertime days above 90 degrees Fahrenheit [32.2°C] is one of the indicators of conditions favorable to ozone formation.

Despite improving air quality and downward trends in most criteria pollutants, ozone will become more challenging to grapple with in a warmer atmosphere. Even if further reductions in air pollution emissions are achieved, what happens if global warming brings hotter and dryer summers over much of the United States? How might such warmer temperatures alter ozone air pollution—the most weather-sensitive pollutant— and what are the resulting health risks?

CLIMATE CHANGE EFFECTS ON AIR POLLUTION

Theoretically, global warming has the potential to increase average ambient concentrations and the frequency of episodes of ozone pollution. As stated in a health report of the U.S. National Assessment on Climate Change, climate change may affect people's exposure to air pollutants by affecting weather, anthropogenic emissions (human-caused), and biogenic emissions (caused by living organisms), and by changing the distribution and types of airborne allergens. Local temperature, precipitation, clouds, atmospheric water vapor, wind speed, and wind direction influence atmospheric chemical processes, and interactions occur between local and global-scale environments.

If the climate becomes warmer and more variable, air quality is likely to be affected for three reasons.

First, weather influences the dispersal and ambient concentrations of many pollutants, in addition to its effects on chemical reaction rates. For example, large high-pressure systems often create a temperature inversion, trapping pollutants in the boundary layer at the earth's surface. It has been suggested that with climate change we are likely to see more instances of very hot weather combined with increases in air-pollutant concentrations.

Second, higher temperatures increase ozone formation when precursors are present. This relationship is nonlinear, with a stronger correlation seen at temperatures above 32 degrees Celsius (90 degrees Fahrenheit). Ozone and nonvolatile secondary PM will generally increase at higher temperatures because of increased gas-phase reaction rates. Interannual temperature variability in California, for example, can increase peak O_3 and 24-hour average PM 2.5 by 16 percent and 25 percent, respectively, when other meteorological variables and emissions patterns are held constant.

Third, woody plants, especially, emit important biogenic volatile organic compounds called *isoprenes*. Isoprene production is controlled primarily by leaf temperature and light. And biogenic VOC emissions are so sensitive to temperature that an increase of as little as 2 degrees Celsius (or 3.6 degrees Fahrenheit) could cause a 25 percent increase in emissions. Therefore, temperature increases could raise ozone levels, in part from vegetative emissions, by increasing the concentration of precursor compounds.

Recent Observations of the Climate-Ozone Relationship

In the eastern United States and in Europe, most days when the levels of ozone exceed air quality standards occur in conjunction with slow-moving high-pressure systems around the summer solstice. This is the period of greatest sunlight, when solar radiation is most intense and air temperatures are high. Episodes of high levels of ozone last three to four days on average and extend over a large area—greater than 600,000 square kilometers (or roughly 230,000 square miles).

During 1995 and 1998, the United States saw relatively high levels of ozone, probably due in part to hot, dry, and stagnant conditions. But in 2003, the United States experienced one of the cleanest ozone years in recent history. The reason had much to do with the unusually cool and wet summer in the eastern United States.

International Reporting

In 1992, under the United Nations Framework Convention on Climate Change (UNFCCC), the United States, along with 185 other countries, agreed to develop and submit a national inventory of anthropogenic greenhouse gas emissions and sinks. To fulfill this obligation, each year the U.S. Environmental Protection Agency (EPA) prepares the official *Inventory of U.S. Greenhouse Gas Emissions and Sinks* in cooperation with the U.S. Department of State and other U.S. government agencies. Under the direction of the Intergovernmental Panel on Climate Change (IPCC), hundreds of scientists and national experts collaborated in developing a set of methodologies and guidelines to help countries create inventories that are comparable across international borders. The information presented in the *Inventory of U.S. Greenhouse Gas Emissions and Sinks* is in full compliance with these IPCC guidelines.

WHAT IS THE SIGNIFICANCE OF EMISSION INVENTORIES?

Greenhouse gas emission inventories are developed for a variety of reasons. Scientists use inventories of natural and anthropogenic emissions as tools when developing atmospheric models. Policy makers use inventories to develop strategies and policies for emissions reductions and to track the progress of those policies. And, regulatory agencies and corporations rely on inventories to establish compliance records with allowable emission rates. Businesses, the public, and other interest groups use inventories to better understand the sources and trends in emissions. A well constructed inventory is consistently prepared, accurate, and thoroughly documented. Inventories typically include the following information:

- Chemical and physical identity of the pollutants

- Types of activities that cause emissions

- Time period over which the emissions were estimated

- Geographic area covered

- Clear description of estimation methodologies used and data collected.

The *Inventory of U.S. Greenhouse Gas Emissions and Sinks* provides important information about greenhouse gases, quantifies how much of each gas was emitted into the atmosphere, and describes some of the effects of these emissions on the environment.

THE U.S. GREENHOUSE GAS INVENTORY PROGRAM

The U.S. Environmental Protection Agency's Clean Air Markets Division (CAMD) in the Office of Atmospheric Programs is responsible for developing the annual *Inventory of U.S. Greenhouse Gas Emissions and Sinks*. EPA's Greenhouse Gas Inventory Program has developed extensive technical expertise, internationally recognized analytical methodologies, and one of the most rigorous management systems in the world for the estimation, documentation, and evaluation of greenhouse gas emissions and sinks for all source categories. To accomplish its work, the Inventory Program collaborates with hundreds of experts representing more than a dozen federal agencies, many academic institutions, industry associations, consultants, and environmental organizations. The Program also works directly with industries and other government agencies to develop high quality emissions data and is supported by CAMD's experience with the U.S. emissions trading programs and its network of continuous emission monitors for CO_2 on most electric power plants in the United States.

WHAT ARE GREENHOUSE GASES?

Many chemical compounds found in the earth's atmosphere act as greenhouse gases, trapping outgoing terrestrial radiation and warming the earth's atmosphere. Some emissions of greenhouse gases occur naturally, while others result from human activities. Carbon dioxide, methane, nitrous oxide, and ozone are greenhouse gases that have both natural and human-related emission sources. In addition, humans have created other greenhouse gases, such as hydrofluorocarbons (HFCs), perfluorocarbons (PFCs), and sulfur hexafluoride (SF_6). The global warming potential (GWP) of a greenhouse gas is the ratio of global warming, or radiative forcing, from the emission of one unit mass of a greenhouse gas to that of one unit mass of carbon dioxide over a specified time horizon. Calculation of GWPs is based on the lifetime of the gas and how efficiently it traps heat in the atmosphere.

WHAT IS THE *INVENTORY OF U.S. GREENHOUSE GAS EMISSIONS AND SINKS*?

The *Inventory of U.S. Greenhouse Gas Emissions and Sinks* is a catalog of anthropogenic, or human-generated, greenhouse gas emissions in the United States. Carbon dioxide can also be sequestered (i.e., stored) in "sinks" that result from forestry and other land-use practices. Excluding all naturally occurring greenhouse gas emissions and sinks, the *Inventory* provides a detailed record of all emissions and sinks directly attributable to human activities. It does not address naturally occurring emissions or sinks.

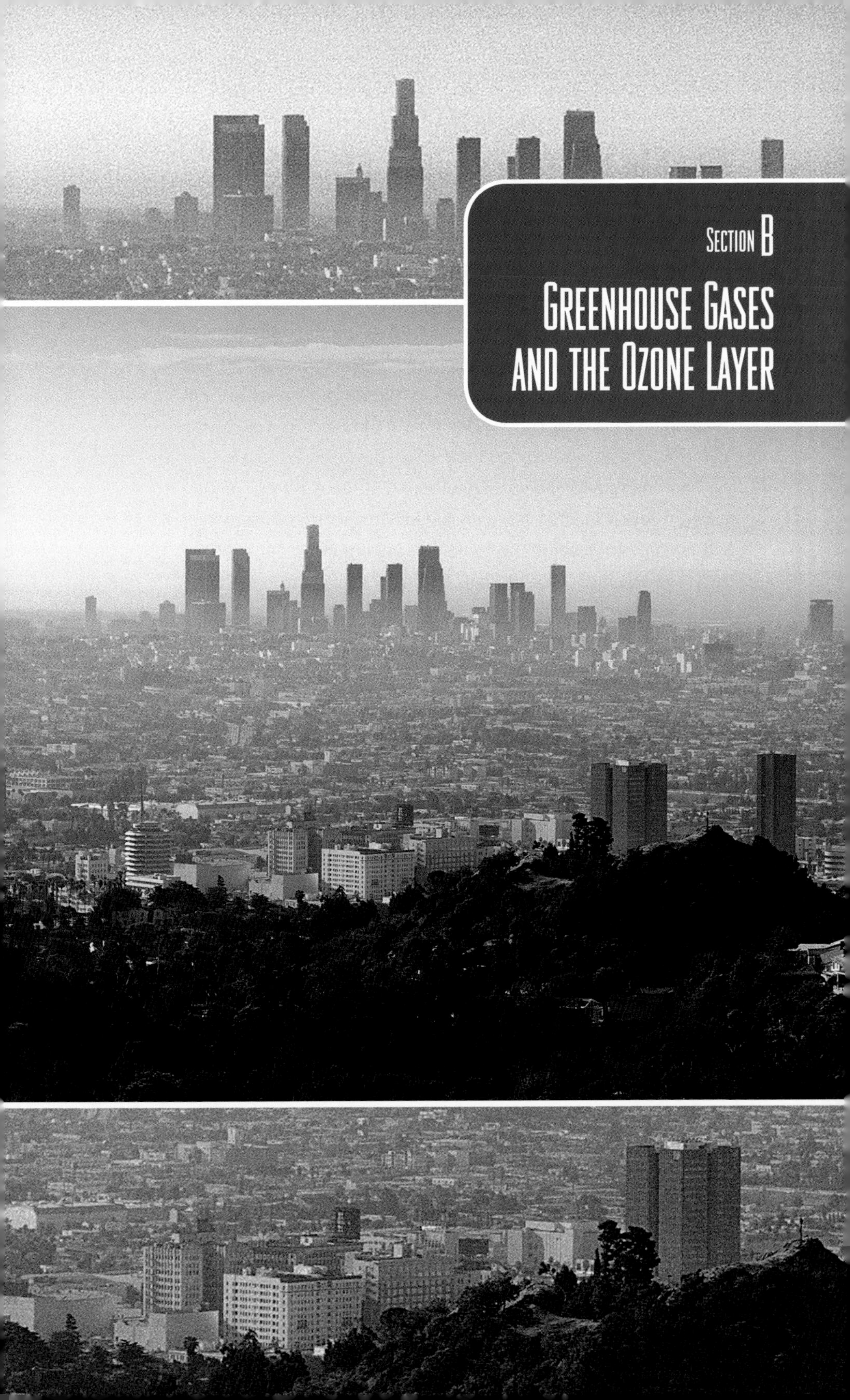

Section B

Greenhouse Gases and the Ozone Layer

What Is the Greenhouse Effect?

The burning of fossil fuels (gas, oil, and coal) is a key to the economies of many industrialized nations. But the gases and particulate matter released when they are burned contribute to global warming. Most scientists who study the issues agree that if the recent rates of greenhouse gas (GHG) emissions are not reduced, there will be serious damage to the Earth's ecosystems and the humans who rely on them.[1]

The Intergovernmental Panel on Climate Change (IPCC) issued a report in 2001 that concluded: "In light of new evidence and taking into account the remained uncertainties, most of the observed warming over the last 50 years is likely to have been due to the increase in greenhouse gas concentrations."[2]

To help understand what GHGs are and why they cause global warming, the following information from the U.S. Environmental Protection Agency (EPA) is included. Some greenhouse gases include water vapor, carbon dioxide, methane, ground-level ozone, nitrogen oxides, and some volatile organic compounds. As solar radiation passes through our atmosphere, various particles transmit and absorb the radiation. The significance of the GHGs, which make up a very small part of the atmosphere, is that they absorb certain infrared wavelengths and trap heat in the lower atmosphere in almost the same way that a closed car sitting in the sun heats up.[3]

Each year, the EPA and other government agencies complete an emissions inventory, called the *Inventory of U.S. Greenhouse Gas Emissions and Sinks*, to identify and to quantify our emissions of greenhouse gases. The second article explains the program that provides important information about climate change to policymakers. A complete copy of the most recent report is available the EPA Website.[4]

—The Editor

1. Hardy, John T. *Climate Change: Causes, Effects, and Solutions*. New York: John Wiley and Sons, 2003, p. 211.

2. Intergovernmental Panel on Climate Change (IPCC). Third Annual Report: Climate Change 2001: A Scientific Basis. (Section E.8) Available online at *http://www.grida.no/climate/ipcc_tar/wg1/028.htm#e8*.

3. Hardy, p. 7.

4. *Inventory of U.S. Greenhouse Gas Emissions and Sinks: 1990–2002*. U.S. Environmental Protection Agency, Office of Atmospheric Programs, EPA 430-R-04-003, April 2004.

Emissions

from the U.S. Environmental Protection Agency (EPA)

Once, all climate changes occurred naturally. However, during the Industrial Revolution, we began altering our climate and environment through changing agricultural and industrial practices. Before the Industrial Revolution, human activity released very few gases into the atmosphere, but now through population growth, fossil fuel burning, and deforestation, we are affecting the mixture of gases in the atmosphere.

What Are Greenhouse Gases?

Some greenhouse gases occur naturally in the atmosphere, while others result from human activities. Naturally occurring greenhouse gases include water vapor, carbon dioxide, methane, nitrous oxide, and ozone. Certain human activities, however, add to the levels of most of these naturally occurring gases:

Carbon dioxide is released to the atmosphere when solid waste, fossil fuels (oil, natural gas, and coal), and wood and wood products are burned.

Methane is emitted during the production and transport of coal, natural gas, and oil. Methane emissions also result from the decomposition of organic wastes in municipal solid waste landfills, and the raising of livestock.

Nitrous oxide is emitted during agricultural and industrial activities, as well as during combustion of solid waste and fossil fuels.

Very powerful greenhouse gases that are not naturally occurring include *hydrofluorocarbons* (HFCs), *perfluorocarbons* (PFCs), and *sulfur hexafluoride* (SF_6), which are generated in a variety of industrial processes.

Each greenhouse gas differs in its ability to absorb heat in the atmosphere. HFCs and PFCs are the most heat-absorbent. Methane traps over 21 times more heat per molecule than carbon dioxide, and nitrous oxide absorbs 270 times more heat per molecule than carbon dioxide. Often, estimates of greenhouse gas emissions are presented in units of millions of metric tons of carbon equivalents (MMTCE), which weights each gas by its GWP value, or Global Warming Potential.

What Are Emissions Inventories?

An emission inventory is an accounting of the amount of air pollutants discharged into the atmosphere. It is generally characterized by the following factors:

- the chemical or physical identity of the pollutants included,

- the geographic area covered,

- the institutional entities covered,

- the time period over which emissions are estimated, and

- the types of activities that cause emissions.

Emission inventories are developed for a variety of purposes. Inventories of natural and anthropogenic emissions are used by scientists as inputs to air quality models, by policy makers to develop strategies and policies or track progress of standards, and by facilities and regulatory agencies to establish compliance records with allowable emission rates. A well constructed inventory should include enough documentation and other data to allow readers to understand the underlying assumptions and to reconstruct the calculations for each of the estimates included.

What Are Sinks?

A sink is a reservoir that uptakes a chemical element or compound from another part of its cycle. For example, soil and trees tend to act as natural sinks for carbon—each year hundreds of billions of tons of carbon in the form of CO_2 are absorbed by oceans, soils, and trees.

What Is the Ozone Hole?

In the early 1970s, at the University of California, two scientists published a paper that identified chlorofluorocarbons (CFCs) as a very serious threat to our stratospheric ozone layer. The ozone layer rests about 10 to 20 miles [16 to 32 km] above the Earth's surface and absorbs the sun's harmful radiation (ultraviolet, or UV, light). In 1995, these two scientists, Sherwood Rowland and Mario Molina, won the Nobel Prize for Chemistry for their research on the human threat to the ozone layer.[1]

Ozone that is located close to the Earth in the tropospheric layer threatens the health of plants and animals. Ozone that is located in the stratosphere protects us. How can that be when it is the same chemical substance—three molecules of oxygen bonded together? It is, in fact, only location that determines whether ozone will protect us from the harmful rays of the sun or cause health problems.

Attempts at some sort of international cooperation to deal with the ozone problem failed until 1987. The Montreal Protocol of that year was the first international agreement to address the issue, and member countries agreed to reduce the use of such ozone-destroying chemicals as CFCs and bromine-containing halons. The Protocol has been revised five times since its creation, each time making the controls more stringent.[2] The reduction of ozone-depleting chemicals has been largely successful. The process and its results provide hope for countries working together on global pollution issues.

However, the problem of the thinning ozone layer has not stopped. A 2003 press release from the National Oceanic and Atmospheric Administration (NOAA) stated that the ozone hole over the Antarctic is near record size. Part of the problem is that the chemicals that destroy stratospheric ozone remain active in the atmosphere for about 50 years. Thus, the current Antarctic ozone hole may take 50 years to repair.[3]

The following information on stratospheric ozone and the hole in the ozone layer comes from the National Oceanic and Atmospheric Administration (NOAA). NOAA was established in

1970, but some of the agencies that formed it were among the oldest in our government, some dating from 1807. NOAA does global research on oceans, atmosphere, space, and the sun. Its Website contains information about its various programs: the national weather service, NOAA fisheries, Marine Sanctuaries, and many others.

—The Editor

1. UCS backgrounder. *The Science of Stratospheric Ozone Depletion*, 2002.

2. Parson, Edward A. *Protecting the Ozone Layer: Science and Strategy.* New York: Oxford University Press, 2003, p. 4.

3. NOAA Press Release. *2003 Antarctic Ozone "Hole" Near Record Size.* October 7, 2003. Available online at *http://www.ozonelayer.noaa.gov/ press/press.htm.*

National Oceanic and Atmospheric Administration (NOAA) Factsheets
from the National Oceanic and Atmospheric Administration

OZONE BASICS

Ozone is very rare in our atmosphere, averaging about three molecules of ozone for every 10 million air molecules. In spite of this small amount, ozone plays a vital role in the atmosphere. In the information below, we present "the basics" about this important component of the Earth's atmosphere.

Where Is Ozone Found in the Atmosphere?

Ozone is mainly found in two regions of the Earth's atmosphere. Most ozone (about 90%) resides in a layer that begins between 6 and 10 miles (10 and 17 kilometers) above the Earth's surface and extends up to about 30 miles (50 kilometers). This region of the atmosphere is called the stratosphere. The ozone in this region is commonly known as the ozone layer. The remaining ozone is in the lower region of the atmosphere, which is commonly called the troposphere. . . .

What Roles Does Ozone Play in the Atmosphere and How Are Humans Affected?

The ozone molecules in the upper atmosphere (stratosphere) and the lower atmosphere (troposphere) are chemically identical, because they all consist of three oxygen atoms and have the chemical formula O_3. However, they have very different roles in the atmosphere and very different effects on humans and other living beings. Stratospheric ozone (sometimes referred to as "good ozone") plays a beneficial role by absorbing most of the biologically damaging ultraviolet sunlight (called UV-B), allowing only a small amount to reach the Earth's surface. The absorption of ultraviolet radiation by ozone creates a source of heat, which actually forms the stratosphere itself (a region in which the temperature rises as one goes to higher altitudes). Ozone thus plays a key role in the temperature structure of the Earth's atmosphere. Without the filtering action of the ozone layer, more of the Sun's UV-B radiation would penetrate the atmosphere and would reach the Earth's surface. Many experimental studies of plants and animals and clinical studies of humans have shown the harmful effects of excessive exposure to UV-B radiation.

At the Earth's surface, ozone comes into direct contact with life-forms and displays its destructive side (hence, it is often called "bad ozone"). Because ozone reacts strongly with other molecules, high levels of ozone are toxic to living systems. Several studies have documented the harmful effects of ozone on crop production, forest growth, and human health. The substantial negative effects of surface-level tropospheric ozone from this direct toxicity contrast with the benefits of the additional filtering of UV-B radiation that it provides.

What Are the Environmental Issues Associated With Ozone?

The dual role of ozone leads to two separate environmental issues. There is concern about increases in ozone in the troposphere. Near-surface ozone is a key component of photochemical "smog," a familiar problem in the atmosphere of many cities around the world. Higher amounts of surface-level ozone are increasingly being observed in rural areas as well.

There is also widespread scientific and public interest and concern about losses of ozone in the stratosphere. Ground-based and satellite instruments have measured decreases in the amount of stratospheric ozone in our atmosphere. Over some parts of Antarctica, up to 60% of the total overhead amount of ozone (known as the column ozone) is depleted during Antarctic spring (September–November). This phenomenon is known as the Antarctic ozone hole. In the Arctic polar regions, similar processes occur that have also led to significant chemical depletion of the column ozone during late winter and spring in 7 out of the last 11 years [as of 1999]. The ozone loss from January through late March has been typically 20–25%, and shorter-period losses have been higher, depending on the meteorological conditions encountered in the Arctic stratosphere. Smaller, but still significant, stratospheric decreases have been seen at other, more-populated regions of the Earth. Increases in surface UV-B radiation have been observed in association with local decreases in stratospheric ozone, from both ground-based and satellite-borne instruments.

What Human Activities Affect Upper-Atmospheric Ozone (The Stratospheric Ozone Layer)?

The scientific evidence, accumulated over more than two decades of study by the international research community, has shown that human-produced chemicals are responsible for the observed depletions of the ozone layer. The ozone-depleting compounds contain various combinations of the chemical elements chlorine, fluorine, bromine, carbon, and hydrogen and are often described by the general term halocarbons. The compounds that contain only chlorine, fluorine, and carbon are called chlorofluorocarbons, usually abbreviated as CFCs. CFCs, carbon tetrachloride, and methyl chloroform are important human-produced ozone-depleting gases that have been used in many applications including refrigeration, air conditioning, foam blowing, cleaning of electronics components, and as solvents. Another important group of human-produced halocarbons is the halons, which contain carbon, bromine, fluorine, and (in some cases) chlorine and have been mainly used as fire extinguishants.

What Actions Have Been Taken to Protect the Ozone Layer?

Through an international agreement known as the Montreal Protocol on Substances that Deplete the Ozone Layer, governments have decided to eventually discontinue production of CFCs, halons, carbon tetrachloride, and methyl chloroform (except for a few special uses), and industry has developed more "ozone-friendly" substitutes. All other things being equal, and with adherence to the international agreements, the ozone layer is expected to recover over the next 50 years or so.

OZONE DEPLETION

In the stratosphere, the region of the atmosphere between about 6 and 30 miles (10 and 50 kilometers) above the Earth's surface, ozone (O_3) plays a vital role by absorbing harmful ultraviolet radiation from the sun. Stratospheric ozone is threatened by some of the human-made gases that have been released into the atmosphere, including those known as chlorofluorocarbons (CFCs). Once widely used as propellants in spray cans, refrigerants, electronics cleaning agents, and in foam and insulating products, the CFCs had been hailed as the "wonder chemicals." But the very properties that make them useful—chemical inertness, non-toxicity, insolubility in water—also make them resistant to removal in the lower atmosphere.

CFCs are mixed worldwide by the large-scale motions of the atmosphere and survive until, after 1–2 years, they reach the stratosphere and are broken down by ultraviolet radiation. The chlorine atoms within them are released and directly attack ozone. In the process of destroying ozone, the chlorine atoms are regenerated and begin to attack other ozone molecules . . . and so on, for thousands of cycles before the chlorine atoms are removed from the stratosphere by other processes. . . .

The "ozone layer" resides at an altitude of about 12 to 15 miles (20 to 25 kilometers) above sea level. It acts as a shield by absorbing biologically active ultraviolet light (called UV-B) from the sun. If the ozone layer is depleted, more of this UV-B radiation reaches the surface of the earth. Increased exposure to UV-B has harmful effects on plants and animals, including humans. The chlorine and bromine in human-produced chemicals such as the ones known as chlorofluorocarbons (CFCs) and halons are depleting ozone in the stratosphere. The figure shows a simplified cycle of reactions in which chlorine (Cl) destroys ozone (O_3).

THE ANTARCTIC OZONE HOLE

The Antarctic Ozone Hole was discovered by the British Antarctic Survey from data obtained with a ground-based instrument from a measuring station at Halley Bay, Antarctica, in the 1981–1983 period. They reported the October ozone loss in 1985. Satellite measurements then confirmed that the springtime ozone loss was a continent-wide feature.

Research conducted during the National Ozone Expeditions to the U.S. McMurdo Station in 1986 and 1987, and NASA [National Aeronautics and Space Administration] stratospheric aircraft flights into the Antarctic region from Chile in 1987 showed conclusively that the ozone loss was related to halogen (chlorine)-catalyzed chemical destruction which takes place following spring sunrise in the Antarctic polar region. The chlorine is derived from manmade chlorofluorocarbons (CFCs) which have migrated to the stratosphere and have been broken down by solar ultraviolet light, freeing chlorine atoms.

The ozone hole is formed each year in the Southern Hemisphere spring (September–November) when there is a sharp decline (currently up to 60%) in the total ozone over most of Antarctica. During the cold dark Antarctic winter, stratospheric ice clouds (PSCs, polar stratospheric clouds) form when temperatures drop below -78°C [-108.4°F]. These clouds are responsible for chemical changes that promote production of chemically active chlorine and bromine. When sunlight returns to the Antarctic in the Southern Hemisphere spring, this chlorine and bromine activation leads to rapid ozone loss, which then results in the Antarctic ozone hole. Although some ozone depletion also occurs in the Arctic during the Northern Hemisphere spring (March–May), wintertime temperatures in the Arctic stratosphere are not persistently low for as many weeks which results in less ozone depletion.

Owing to regulations on the production and use of certain ozone-destroying chlorinated compounds, which went into effect in January 1996, the atmospheric concentration of some of these man-made substances has begun to decline. Chlorine/bromine should reach maximum levels in the stratosphere in the first few years of the 21st century, and ozone concentrations should correspondingly be at their minimum levels during that time period. It is anticipated that the recovery of the Antarctic Ozone Hole can then begin. But because of the slow rate of healing, it is expected that the beginning of this recovery will not be conclusively detected for a decade or more, and that complete recovery of the Antarctic ozone layer will not occur until the year 2050 or later. The exact date of recovery will depend on the effectiveness of present and future regulations on the emission of CFCs and their replacements. It will also depend on climate change in the intervening years, such as long-term cooling in the stratosphere, which could exacerbate ozone loss and prolong recovery of the ozone layer.

Although increasing greenhouse gas concentrations in the atmosphere may result in warmer surface temperatures, colder temperatures are expected to occur in the stratosphere. In fact, temperatures in the lower stratosphere, as measured by NOAA's Microwave Sounding Unit, have cooled during the past 22 years, the length of the satellite record. Colder stratospheric temperatures can enhance ozone loss through their affect on the formation of polar stratospheric clouds which in turn promote chlorine-caused ozone destruction.

Will Climate Change Affect Our Nation's Soils?

Do power plants that burn coal and emit carbon dioxide have anything to do with wheat farms in the Midwest? Yes, because the process of burning coal gives off carbon dioxide, a greenhouse gas that contributes to rising global temperatures. Higher temperatures can change the way rain falls in the world. Rainfall affects crop growth and rates of soil erosion. The Intergovernmental Panel on Climate Change (IPCC) stated in its 2002 report *Climate Change and Biodiversity* that: "Increasing global mean surface temperature is very likely to lead to changes in precipitation and atmospheric moisture because of changes in atmospheric circulation, a more active hydrological cycle. . . ."[1]

The following article from the *Journal of Soil and Water Conservation* discusses a study by a panel of experts in climatology, hydrology, and soil erosion. The study found that changes in precipitation patterns caused by global warming could increase the problems of eroding farmlands. With additional soil erosion come increased water pollution and the loss of valuable farmland.

Just as some international conservation organizations are developing plans to help people and animals deal with the effects of climate change, soil conservationists in the United States must "adapt their conservation planning procedures and tools to a climate that poses greater risks of erosion, runoff, and nonpoint source pollution." Part of the strategy must include starting programs to conserve topsoil in different regions of the country where rainfall patterns become more intense.

—The Editor

1. Intergovernmental Panel on Climate Change (IPCC). *Climate Change and Biodiversity*. IPCC Technical Paper V, 2002.

As the Planet Heats Up, Will Topsoil Melt Away?

by Paul D. Thacker

The most shocking finding was that a relatively small increase in rain intensity, 10 percent, which we've already seen in many areas in the last forty to fifty years means an average 24 percent increase in soil erosion.

—Jim Bruce, the Canadian policy representative for the Soil and Water Conservation Society

This last summer [2003] was one of the worst growing seasons in over a decade for farmers in the Midwest. With little rain, crops roasted in the summer sun, turning ears of corn into shriveled nubs. Then in a cruel twist of irony, late summer rains appeared, too late for corn, which had been cut for silage, but the thunderstorms did cause flooding and a handful of drownings that made national press. What didn't make the evening news is that scientists predict that because of global warming this summer's erratic rainfall patterns and high intensity storms will become the norm in future growing seasons.

Late last year, the Soil and Water Conservation Society brought together a panel of experts to study how changing rainfall will affect soil erosion on cropland. Drawing from a range of expertise—climatologists, hydrologists, and soil erosion experts—they released a report that found climate change will cause rainfall to increase across much of the United States while bringing about desertification to other states. Most importantly, global warming will lead to more frequent disastrous rains, compounding the effects of soil erosion.

To understand rainfall events, the group attacked the issue from two angles. First, a team lead by Pavel Groisman, a climatologist from the University Corporation for Atmospheric Research, analyzed rainfall data from the National

Climate Center. "You cannot say that at a certain farm you see any trend, but when you look at regions, such as the Midwest, you can find signals."

MORE INTENSE RAINFALL

To be reliable, Groisman accessed rainfall data going back to the 1890s. He then ran multiple data points through statistical analysis to find changes in precipitation and to remove possible data errors. What he found was both increased average rainfall over the last century, as well as, increased frequency of intense storms.

More importantly, his data show a higher rate of increase for both average rainfall and rainfall intensity during the last thirty years. "It's a new phenomenon in the history of rainfall," he adds.

Groisman's work was then compared with future projections from two global climate models, one developed by the Canadian Centre for Climate Modeling and Analysis and the other by the Hadley Centre for Climate Prediction and Research. Both models found that average rainfall values will generally increase, but disagreed on specific numbers. However, Groisman adds that when the models compared high rain events (upper five percent of intensity), they started to agree with each other. These recent trends are consistent with climate change projections.

It is these extreme thunderstorm events that both surprise and worry scientists. When rainfall numbers were run through erosion algorithms, scientists found high numbers for future levels of erosion. "The most shocking finding was that a relatively small increase in rain intensity, 10 percent, which we've already seen in many areas in the last forty to fifty years, means an average 24 percent increase in soil erosion," says Jim Bruce, the Canadian policy representative for the Soil and Water Conservation Society. "It's a really big magnification factor."

Between 1985 and 1995, cropland erosion was reduced by almost 40 percent. Today, however, more than 26 million hectares (65 million acres) of farmland erode at rates that threaten agricultural productivity. But decreased rainfall might also lead to erosion.

"Anywhere you have an increase in annual rainfall or intensity, we can expect more erosion," says Mark Nearing, a soil scientist with the USDA. "And in parts of the United States where there will be decreases in precipitation such as the Great Plains, we can still find an increase in erosion because of loss of plant biomass."

CORRALLING A FUTURE THUNDERSTORM

Trying to spot very specific future trends in rainfall is simply not possible at this time as the global climate models are still rather crude. Basic trends can be found but in areas such as the western United States, the Sierra Mountain Range plays havoc on the models. Still, the experts agree that agricultural policy must evolve and keep careful watch on future rainfall patterns.

Besides rethinking how to control erosion runoff, and perhaps beefing up engineering structures that slow topsoil loss, Jerry Hatfield, laboratory director of the National Soil Tilth Lab, says we have to rethink many different management practices. In the case of manure application, he noted that extreme rainfall might lead to more water pollution.

"Current practices with manure application were developed over the last fifteen to twenty years and many of our assumptions were developed under times when the climate was reasonably stable," he says. "That's not happening now." Hatfield doesn't expect manure application to change dramatically, but he does feel the need to "tweak the system to keep moving along." He also adds that conservation tillage, which calls for 30 percent biomass cover on the ground, might need to leave more plant material behind to help slow topsoil loss. "If you look at the increased rain intensity and increased rainfall, then maybe 30 percent residue cover might not be sufficient."

Bruce also says that excess erosion could lead to a greater problem with pollution and herbicide runoff. "It not only has a big impact on farming if you lose topsoil to erosion, but we can get more nutrients washing into water bodies such as Lake Erie and Lake Ontario," he says. "And with big factory farms and manure piles we can have *E. coli* running into our water supplies. So we're going to have to be more vigilant, so we don't have disease outbreaks."

The best way to keep farmers' topsoil safe on the farm seems to be rethinking not just farming practices, but federal programs and funding. The last farm bill contained $17 billion targeted at conservation, but conservation agencies will need to adapt their conservation planning procedures and tools to a climate that poses greater risks of erosion, runoff, and nonpoint source pollution.

Unlike the United States, Canada does not have a national program for soil erosion. But Bruce says that each province in Canada must become more aggressive to limit erosion increases during future growing seasons. As for American scientists, they say that federal monies will need to be shifted around to different parts of the country, perhaps being concentrated more in areas where rainfall intensities become the greatest. Everyone feels that government programs are not doing enough.

"Farmers are the people most interested in this because they are seeing problems such as fields being washed out that didn't use to erode so heavily," says Nearing. "They're not happy about it."

In the coming years, farmers complaining about washed out fields and poor harvests due to erosion will probably become more and more a part of the news. Scientists are now predicting that this will occur, but will government action keep topsoil on the land?

"Our agency needs to gear up and understand how to deal with this," says Nearing. "We haven't even started to get prepared."

Will Climate Change Affect the Great Lakes Region?

Over the past 50 years, human activities in the industrialized world have contributed greatly to global warming. In 2001, the Intergovernmental Panel on Climate Change (IPCC) issued its Third Annual Report, giving the opinion that "Although some regions may experience beneficial effects of climate change (e.g., increasing agricultural productivity at high latitudes), previous IPCC assessments have concluded that net negative climate impacts are more likely in most parts of the world."[1]

Scientists have looked at patterns over the Great Lakes region and observed climate changes that include more frequent and severe rainstorms and shorter winters. If this trend continues, lake water levels and nutrient circulation patterns will change. Streams and wetlands will be subject to changes in streamflows, which will affect amphibians and migrating birds. River flooding may increase. The species of trees growing in the forests will change with the warmer weather. Such changes will impact the wildlife in the area. Agriculture may be more susceptible to weather extremes. In addition, the potential for health risks to humans will increase.

The following report from the Union of Concerned Scientists and the Ecological Society of America emphasizes to individuals and policymakers how important it is to plan for future change. It advises that the time to get ready for the impacts of climate change is now, and that we can make the necessary adaptations.

—The Editor

1. Intergovernmental Panel on Climate Change (IPCC). Third Annual Report. *Climate Change 2001: Impacts, Adaptations and Vulnerability*. Section 1.2. Available online at *http://www.grida.no/climate/ipcc_tar/wg2/054.htm*.

2. Union of Concerned Scientists and the Ecological Society of America. *Confronting Climate Change in the Great Lakes Region.* 2002. Available online at *http://www.ucsusa.org/greatlakes/glchallengetoc.html*.

3. Union of Concerned Scientists Website. Available online at *http://www.ucsusa.org/ucs/about/index.cfm*.

Trouble in the Heartland
by Susan Moser

Change is in the air, but it's not just the change of seasons. A new report released in April by the Union of Concerned Scientists and the Ecological Society of America, *Confronting Climate Change in the Great Lakes Region: Impacts on Our Communities and Ecosystems*, shows a profound and potentially threatening change in climate is underway. So profound, indeed, that it may alter the face of North America's heartland as we know it.

Scientists now agree that human activities, such as using electricity produced by coal or driving conventional gasoline-powered cars, are a major contributor to the increase of heat-trapping gases in our atmosphere. Over the past 50 years, these human activities have been the major contributor to rising temperatures and other changes observed around the world. Data collected in the Great Lakes region of the United States and Canada confirm these global trends:

- average annual temperatures are increasing;

- severe rainstorms have become more frequent;

- winters are getting shorter; and,

- the duration of lake ice cover is decreasing as air and water temperatures rise.

PROJECTING THE FUTURE

New climate modeling results used in *Confronting Climate Change in the Great Lakes Region* suggest that the region's climate will grow warmer and probably drier during the 21st century. Climate models predict that by the end of century, temperatures could rise 5 to 12°F (3 to 7°C) in winter and 5 to 20°F (3 to 11°C) in summer. Nighttime temperatures are likely to warm more than daytime temperatures, and extreme heat will be more common. This dramatic warming over the next 100 years is on the same order of magnitude as the warming since the last ice age.

Scientists also expect the seasonal distribution of precipitation in the region to vary greatly, increasing in winter and decreasing in summer. While annual average precipitation may not change much, the region is likely to grow drier overall because rain and snow can't compensate for the drying effects of a warmer climate. Thus, we could well see lower lake levels, drier soils, and perhaps even more droughts, punctuated by more intense rainstorms and flooding.

These changes will dramatically affect how the climate feels to us. In less than three decades, for example, a summer in Illinois may well feel like a summer in Oklahoma today. By the end of the century, an Illinois summer may feel like one in east Texas today. What will these changes mean for Great Lakes ecosystems and the people who live in the region?

LAKES, STREAMS, AND FISHES

Levels of both small inland lakes and the Great Lakes, while always variable, are likely to decline overall in the future, while water temperatures continue to rise. So, although fewer fish may die off in the winter, suitable habitat for coldwater fish such as lake trout or coolwater fish such as walleye and yellow perch will probably shrink. Habitat for warm-water species such as smallmouth bass and bluegill will expand northward, but these fish will face more competition from invasive aquatic species such as the common carp. In addition, more fish could die off in the summer due to oxygen depletion in the warmer water. Lower lake levels, warmer water, and less oxygen also increase problems with mercury contamination—a health threat recreational and commercial fishers already face.

WETLANDS, AMPHIBIANS, SHOREBIRDS, AND WATERFOWL

Earlier ice breakup and spring runoff, more intense flooding, and lower summer water levels mean trouble for wetlands and the species that depend on them. Habitat will dry up. Frogs and salamanders will find fewer places to breed, live, and hide from damaging ultraviolet radiation. Migratory birds such as canvasbacks and some warblers that need shorelines and wetlands for food, resting, and nesting places will find less suitable habitat. Unfortunately, agriculture and development prevent many wetlands from moving north to cooler, wetter climes.

FORESTS AND FORESTRY

The future of Great Lakes forests—now dominant mostly in the northern parts of the region—is harder to predict. Northern forests of blue spruce, fir, and hemlock are likely to shrink. Other tree species will attempt to move northward, but only if they can keep up with the pace of climate change, can find new habitat to colonize, and don't run up against barriers such as human development or unsuitable soils. Warmer temperatures and higher levels of carbon dioxide (CO_2) and nitrogen in the atmosphere could stimulate tree growth, but higher concentrations of ground level ozone, more frequent droughts and forest fires, and a greater risk from insect pests such as the gypsy moth could damage long-term forest health and productivity.

AGRICULTURE

Fields of wheat, corn, and soybeans, dairy farms, and fruit orchards are as much a part of the Great Lakes landscape as its forests, wetlands, and lakes. And like the forests, how agriculture will fare in a warmer world is still uncertain.

Moderately warmer temperatures, higher CO_2 and nitrogen concentrations, and a longer growing season are likely to help increase productivity. But higher ozone levels and changes in the water cycle—more intense rainfall and possibly more flooding during planting and harvesting time, with less moisture available during the main growing season—will make farming significantly harder to manage, especially for smaller farmers. Warmer, wetter winters will also produce more pests and can negatively affect fruit trees, which depend on hard frosts to strengthen tree growth and set buds.

HUMAN HEALTH

With average temperatures increasing, summer heat waves and "bad ozone days" will become more frequent. Scientists expect the number of deaths due to air pollution and heat-related illnesses to increase, especially in the region's many large cities (Chicago, Cleveland, Detroit, Milwaukee, Minneapolis–St. Paul, and Toronto). In the Toronto-Niagara region, for example, the current number of days with maximum temperatures above 86°F (30°C) could double by the 2030s and surpass 50 days by the 2080s. On the other hand, bitterly cold winters will become less common.

Other health-related problems related to global warming include the spread of waterborne diseases such as *cryptosporidiosis*, which struck Milwaukee in 1993 with deadly consequences. Illnesses such as St. Louis encephalitis and Lyme disease, which are transmitted by insects, will likely expand their geographic ranges because the insects are more likely to survive milder winters and reproduce more rapidly as temperatures rise. Whether the increased risk from these threats actually translates into more cases or even deaths depends on the responsiveness of health care systems and the public's degree of preparedness.

A CORNUCOPIA OF SOLUTIONS

The Great Lakes region has always been the industrial and agricultural heartland of North America. The states' combined economic strength makes the region the third largest economy in the world. This has produced enormous benefits for the region's 60 million residents as well as major changes to the landscape—changes that exert significant pressure on the environment and contribute to global warming. The good news, however, is that the Great Lakes can also become the heartland of solutions to climate change.

First, the region can significantly reduce its emissions of heat-trapping gases by switching to less carbon-intensive power production, increasing development of renewable energy sources such as wind and biomass, improving energy efficiency, and producing more fuel-efficient cars. This transition can actually create jobs, save consumers and communities money, and shape a new, smarter economy.

Second, the region can minimize the pressure on its precious ecosystems and resources by improving air and water quality, urban planning, and habitat protection. Such improvements would have immediate benefits for both people and ecosystems, while making them less vulnerable to the future effects of climate change.

And finally, residents of the region can begin planning how they will manage those environmental changes that are unavoidable. Preparing for climate change while working to minimize its pace and severity does not guarantee a future free from surprises. But such prudent management may help avoid unnecessary suffering and preserve the natural treasures of North America's heartland.

How Will Climate Change Affect the "Protected Areas" of the World?

Protected areas are just that—regions of land or sea that have been dedicated to protecting or maintaining biological diversity and the associated cultural resources. These include areas such as wilderness, national parks, national monuments, and other legally protected or managed sites around the world.[1] Today, designated protected areas cover almost 8% of the Earth's land surface, and about 1% of the oceans. But many of the places are remote or are not in places that protect biodiversity.

The World Wildlife Fund (WWF) has been working for over 40 years to increase the numbers of protected areas around the world. The following WWF 2003 report, *No Place to Hide: Effects of Climate Change on Protected Areas*, addresses the threats of climate change to these protected areas. The threats are the same as everywhere else in the world: rising sea levels, melting glaciers and ice caps, changing patterns of weather, and shifting plant and animal ranges and reproductive cycles.

Because these areas have been selected for protection, the effects of climate change are of special concern. As outlined at the end of the report, research and observations of current trends helps protection efforts in several ways. They help prevent additional climate change, plan for change, and learn about any changes that do take place.

—The Editor

1. WWF Position Paper. *Protected Areas*. 2003. Available online at *http://www.panda.org/downloads/protectedareaspositionpaperwpc2003_svvl.pdf.*

2. WWF Website. Available online at *http://www.panda.org/about_wwf/.*

No Place to Hide: Effects of Climate Change on Protected Areas
from the World Wildlife Fund

For the past decade, WWF has identified the potential threats to biodiversity posed by climate change. We have argued that protected areas offer a limited defence against problems posed by rapid environmental change and that protected areas will themselves need to be changed and adapted if they are to meet the challenges posed by global warming. Recent research suggests that the types of environmental changes predicted in climatic models are now taking place. Studies of many animals and plants that show significant alterations in range or behaviour find that the most consistent explanation for these is climate change.

Global warming is also linked to observed impacts such as coral bleaching and melting glaciers. Such changes are taking place widely and as a result, global warming is already impacting protected areas.

These impacts may necessitate a fundamental rethinking in the approach to protection. Protected areas are rooted in the concept of permanence: protection works best as a conservation tool if the area remains protected for the foreseeable future.

Protected area agencies have rightly resisted attempts to move protected area boundaries or de-designate protected areas for periods of time. But under climate change, species for which a particular protected area was established may no longer survive there. Some protected areas—for instance in coastal, arctic and montane regions—may disappear altogether in their current form. The current extinction crisis is likely to become more intense as environmental conditions change and fluctuate at an unnatural rate.

WWF CLIMATE CHANGE PROGRAMME
Protected area agencies could be faced with the massive task of having to shift protected areas to keep up with moving habitats and ecosystems. Some protected areas may have to

retrench onto higher ground as water rises. The practical difficulties should not be underestimated. Protected areas do not exist in an empty landscape and replacement land and water will often not be available. "Moving" protected areas would have enormous implications for their infrastructure, surrounding human communities and the many businesses associated with parks. Shifting reserves would have cultural implications; societies build powerful emotional bonds to national parks and nature reserves that mean they cannot simply be swapped and replaced lightly.

We are still learning about climate change and there have been relatively few studies of impacts within protected areas to confirm or disprove the modeling exercises and speculation. Ecosystems are often quite resilient but while some climate change problems are likely to be surmountable through management, adaptation or evolution, others are likely to be more intractable.

DISAPPEARANCE OF HABITATS AND ECOSYSTEMS

The most intractable threats come from the total disappearance of habitat. For instance the Intergovernmental Panel on Climate Change [IPCC] confirms that sea level rise is already affecting coastal ecosystems, including coral reefs, mangroves and salt-marshes.

A third of the Blackwater National Wildlife Refuge marshland on Chesapeake Bay, USA, has disappeared since 1938 and the rest of the marsh, which provides winter habitat for many waterfowl species, is expected to be flooded within 25 years. While half the existing loss is thought to be due to extraction from aquifers, the rest is believed to be due to sea-level rise. In Waccassassa Bay State Preserve, in Florida, researchers concluded that cabbage palms and other trees are falling victim to saltwater exposure tied to global sea level rise, exacerbated by drought and a reduction of freshwater flow. Rising seas are said to have flooded 7,500 ha [18,533 acres] of mangroves in the Sundarbans National Park of Bangladesh, although sea-level rise is aggravated by subsidence in the delta. In Vietnam, mangroves are reported to be undergoing species change because of increased salt intrusion, including those in Dao Bach Long VI, a proposed marine protected area.

Several Pacific island states are threatened with total disappearance and two uninhabited islands in the Kiribati chain have already disappeared due to sea level rise. The people of Funafti, in Tuvalu are lobbying to find new homes: salt water intrusion has made groundwater undrinkable and these islands are suffering increasing impacts from hurricanes and heavy seas. When they disappear, not only will a unique culture vanish but the Funafuti Marine Conservation Area will be dramatically altered by loss of surface reefs.

The Great Rann of Kutch in Gujarat is one of the emptiest places in India: the vast area of seasonal salt lakes supports huge populations of flamingos and is the only remaining habitat for 2000 Indian wild asses. The area is likely to become inundated by the sea, thus destroying the habitat and threatening the Wild Ass Sanctuary and the Kachchh Desert Sanctuary.

CATASTROPHIC, PROBABLY LONG-TERM CHANGES TO ECOSYSTEMS

Even if ecosystems do not disappear altogether, they may undergo serious and irreversible changes.

Ocean Temperature and Coral Reefs

In 1998, tropical sea surface temperatures were the highest on record, the culmination of a 50-year trend, and simultaneously coral reefs suffered the most extensive and severe bleaching (loss of symbiotic algae) and death on record. This is believed to be due to global warming. Its wide geographical spread and regional severity are apparently due to a steadily rising baseline of marine temperatures, combined with temperature peaks associated with increasingly frequent El Niño and La Niña events.

Climate change threats are compounded by pollution and over-exploitation. Even those reefs in marine sanctuaries are threatened by global climate change.

The Seychelle Islands are justly famous for their coral reefs and the remote Aldabra Atoll is the largest raised atoll in the world. It supports a huge coral diversity and rare land species like the giant tortoise (*Aldabrachelys elephantina*), vast seabird colonies and important sea turtle breeding beaches. Its status is recognised by its listing as a natural World Heritage site by UNESCO [United Nations Educational, Scientific and Cultural Organization]. Apart from a handful of scientists at a research station, the atoll is visited only by occasional passing yachts: the post [mail] boat takes 12 days from the mainland. Yet despite its remoteness and protected status, the Seychelles suffered a severe coral bleaching event in the late 1990s and a recent assessment by the Seychelles Foundation judged climate change to be the most important threat facing the atoll.

In March 2003, WWF reported that coral bleaching was occurring at all its seven research sites in American Samoa, including within the National Park of American Samoa, Fagatele Bay National Marine Sanctuary and Maloata Bay Community Reserve.

Cold Seas

Warming affects cold seas and polar communities as well. Decreased weight in polar bears, which can lead to decreased adult fitness and reproductive success, in the Hudson Bay area of Canada is attributed to earlier spring ice break-up and consequent loss of two weeks hunting. Loss of sea ice due to climate change is a conservation concern; for polar bears of Wrangel Island Zapovednik, a Russian nature reserve, as well.

Rising temperatures are also affecting Antarctica; for instance Adélie Penguin populations have shrunk 33 per cent in 25 years due to declines in winter sea ice habitat, despite the continent having special protected status through the Protocol on Environmental Protection to the Antarctic Treaty.

The Case of Tropical Montane Cloud Forests

As average temperature increases, optimum habitat for many species will move to higher elevations or higher latitudes.

Where there is no higher ground or where changes are taking place too quickly for ecosystems and species to adjust, local losses or global extinctions will occur. Observations suggest that tropical montane cloud forests are highly at risk due to fewer, higher clouds and warmer temperatures, with serious impacts already underway. Amphibians may also be especially susceptible.

Many amphibious species already appear to be declining, and serious losses have been recorded for example in Australia and Central America.

Climate change associated with an El Niño/Southern Oscillation–related drought in 1986/7 is thought to be a cause of amphibian losses in Monteverde Cloud Forest Reserve, a well-managed protected area in Costa Rica. The golden toad (*Bufo periglenes*) and the harlequin frog (*Atelopus varius*) disappeared, and four other frog and two lizard species suffered population crashes. A detailed survey in a 30 km² [11.6 square miles] study site found that 20 out of 50 frog and toad species disappeared at that time.

CATASTROPHIC, TEMPORARY CHANGES TO ECOSYSTEMS

Other changes may be temporary, although the cross-over between temporary and permanent changes is sometimes hard to judge. Drought is a particularly good example of these types of climate change related catastrophies.

Impacts of Droughts on Wetlands and Other Ecosystems

Increased drought is having a measurable impact on many species. For example, drought related decreases in the density and persistence of Green Leaf Warblers (*Phylloscopus trochiloides*) have been recorded on their wintering grounds in the Western Ghats of south India, including protected areas such as Indira Gandhi National Park. Unusual drought is already undermining other protected areas in India. Keoladeo National Park, near Agra, is a wetland area famous as one of the few wintering sites of the highly endangered Siberian crane (*Grus leucogeranus*). It was created as a duck shoot by manipulating a small existing area of wetland and then later declared a national park and World Heritage site. The park is dependent on fresh input of water during each monsoon. Other demands on water have increased and droughts have become more regular, leading to serious water shortages; in the 2003 season, water levels [were] damagingly low.

In 2002, Australia experienced its worst drought since reliable records began in 1910, coupled with the highest average temperatures in a drought year; this has been linked to climate change and created an unprecedented likelihood of fire. A series of severe forest fires occurred in late 2002 and early 2003. In Kosiosko National Park, New South Wales, over fifty separate fires started in an hour due to lightning strikes from an electric storm after months of drought, creating an unstoppable firestorm that burnt alpine vegetation, which is already stressed through being trapped on mountain tops where average temperature is changing fast.

DRAMATIC CHANGES TO HABITATS AND ECOSYSTEMS

Some of the most intense climate change–related habitat alterations are those that affect glaciers and ice-fields. Loss of glaciers changes the whole nature of a park, although impacts on biodiversity are less well known.

Melting of Icecaps and Glaciers

Glaciers are retreating at an unprecedented rate, changing the entire ecology of mountain habitats. Protected area managers can do nothing to prevent this loss and have to stand by as the ecology changes before their eyes.

In the Hohe Tauern National Park in Austria, the Pasterze Glacier has retreated several hundred metres since the 1970s and a similar pattern is seen throughout the European Alps. Areas are expected to experience extra water flow as the glaciers melt, but then a longer term decline. The glaciers in the Blackfoot-Jackson Glacier Basin of Glacier National Park, Montana, USA, decreased from 21.6 km^2 [8.3 square miles] in 1850 to 7.4 km^2 [2.9 square miles] in 1979. And scientists measuring ice changes in Mount Kilimanjaro National Park, containing Africa's highest mountain, predict that the glaciers will be gone by 2020.

Distributional Changes Towards the Poles and to Higher Altitudes

On the ground observations confirm the theories: species are already shifting their territories because of climate change. Analysis of 143 studies showed a consistent temperature-related shift in species ranging from molluscs to mammals and from grasses to trees. A similar study of 1700 species also confirmed climate change predictions, with average range shifts of 6.1 km/decade [3.8 miles/decade] towards the poles.

Detailed examples illustrate a general trend. Insect dispersal to more favourable areas is a common response to changing climate and for example this has recently been recorded in Japan and the ranges of European and North American butterflies are shifting northwards and upwards.

Changes have been recorded in the height at which several bird species are found in Monteverde. These mirror more widespread shifts in bird ranges towards the poles in Antarctica, North America and in Europe where the spring range of Barnacle Geese (*Branta leucopsis*) has moved north along the Norwegian coast, correlated with a significant increase in days with temperatures above 6°C [42.8°F].

Modeling studies suggested that under likely climate scenarios, cacti currently protected in the Tehuacán-Cuicatlán Biosphere Reserve in Mexico would undergo a drastic population reduction and fall almost entirely outside the reserve area.

Climatic warming observed in the European Alps has been associated with upward movement of some plant species of 1–4 m [3.3–13.1 feet] per decade and loss of some taxa that were restricted to high elevations, posing direct threats to protected areas such as the Swiss National Park.

INDIVIDUAL CHANGES TO SPECIES AND LOCAL FOOD WEBS

Rapid temperature changes affect the seasons, including

shorter winters and variations in season length. For example, this can lead to changes in bud break and bird nest building; and also changes in the growing season, as seen in Europe. Climatic records have been put together with long-term records of flowering and nesting times to record trends. In Britain there are records of flowering time and leaf-break as far back as 1736, which provide solid evidence for climate-related changes. Long term trends towards earlier bird breeding, earlier spring migrant arrival and later autumn departure dates have been observed in North America, along with changes in migratory patterns in Europe. In the UK, the blackcap (*Sylvia atricapilla*) and chiffchaff (*Phylloscopus collybita*) warblers increasingly stay all winter instead of migrating, giving them first choice of nesting sites and disadvantaging migratory warblers.

Climate change may also be causing a mismatch in breeding of Great Tits (*Parus major*) with other species.

Such changes can reduce food availability. In California, a 90 per cent decline in sooty shearwaters (*Puffinus griseus*) from 1987 to 1994 has been linked to a warming of the California Current. In Triangle Island, an ecological reserve off Vancouver Island, the 1.1 million breeding colony of Cassin's auklet (*Ptychoramphus aleuticus*) is declining; probably because of a long-term decline in zooplankton, itself caused by warmer sea temperatures. Salmon running in the Fraser River, British Columbia, which rises in Jasper National Park, have experienced over 50 percent mortality in several runs, with highest losses in years with warm river temperatures, themselves linked with climate change.

Warmer temperatures in Europe have been stimulating an increasing number of frogs and toads to spawn before Christmas. In mild winters this may be an advantage in that the tadpoles have longer to mature, but a cold snap can kill the eggs.

In Lake Constance, a Ramsar site at the border of Germany, Austria and Switzerland, the proportion of long-distance migrant birds decreased and short-distance migrants and residents increased between 1980 and 1992, during a period when winter temperature increased, suggesting that warmer winters may pose a more severe threat to long-distance migrants than to the other bird groups.

EFFECTS OF CLIMATE CHANGE ON PROTECTED AREAS

While some species will adapt their behaviour to new conditions, others will remain trapped in lifecycles that no longer work in the new climatic regime. For example, sea level rise will adversely impact breeding seabirds that nest on island beaches, including those in the Galapagos National Park.

Studies of Protected Area Systems

As data accumulate, researchers are able to make ecosystem analyses of climate change impacts. Climate change, for example, is identified as a third major factor in accelerating problems in the Amazon, which includes several national parks, among them Xingu National Park, along with logging and changes in deforestation patterns.

Attempts are now being made to predict changes for protected area networks. A study of Canada's national parks found that in virtually all climate scenarios, substantial vegetation change will occur in over half the protected areas, with loss of tundra and increase in temperate forests. The authors concluded that: "Climate change represents an unprecedented challenge to Parks Canada and its ability to achieve its conservation mandate. . . . " Important changes relate to hydrology, glacial balance, waning permafrost, increased natural disturbance, shorter ice season and range changes.

Analysis of South African protected areas suggests that increased drought and aridity could lead to huge losses of biodiversity. For example there could be potentially large losses of plant species diversity in semi-arid regions with low landscape heterogeneity.

Richtersveld National Park in the Succulent Karoo, one of seven protected areas in this region, covers 21 per cent of the remaining habitat and has been identified as being highly at risk of damage from climate change. In the United Kingdom, the Institute of Terrestrial Ecology estimates that 10 per cent of nature reserves could be lost (e.g., to climate change-induced habitat degradation) within 30–40 years. In the USA, 7,000 miles [11,265 km] of protected shoreline, including 80 coastal parks such as Assateague Island National Seashore, are at risk from sea-level rise. Studies in Australia identify the need for major changes to protected areas conservation strategies.

ADDRESSING THE PROBLEMS

The predictions that WWF, and others, have been making about the impacts of climate change on protected areas are coming true. Most protected area authorities are still not taking this issue seriously: in many countries the immediate pressures on parks and shortages of resources mean there is no time to worry about future impacts, even if they are already becoming manifest. Clearly, protected area agencies and managers need to consider climate change in future plans. But what exactly should they do? WWF suggests that four urgent steps are needed:

Preventing Change: The optimum way to reduce the impacts of climate change on protected areas is to dramatically reduce the heat-trapping gases that cause climate change. Carbon dioxide, the main heat trapping gas, is emitted by the burning of fossil fuels (coal, oil and natural gas). If these emissions are not cut deeply and quickly, there will be little chance of saving many protected areas. The power sector is responsible for 37% of those emissions globally and has many opportunities to switch from coal to clean power. To learn more go to *www.panda.org/climate*.

Managing for Change: Many climate change impacts are exacerbated by other pressures: even climate-related phenomena like coral bleaching and dieback are increased by pollution and mechanical damage. WWF is publishing a guide to adaptation strategies, "Buying Time: A User's Manual for Building Resistance and Resilience to Climate Change in Natural Systems," the first comprehensive account of how protected areas might be managed in a rapidly changing climate.

Planning for Change: Protected area agencies need advice—and political support—for planning protected area networks to withstand or adapt to change as much as possible. This needs to be a collaborative, global exercise. WWF proposes that the World Commission on Protected Areas would be one obvious body to coordinate such an effort, perhaps as a task force under its theme on management effectiveness.

Learning About Change: We are still in a transition from theory to observation in terms of climate change impacts on protected areas. Information comes from a very limited number of examples. Along with an urgent need to learn more, we also need a series of controlled exercises in addressing and hopefully mitigating pressures when they arise.

WWF CLIMATE CHANGE PROGRAMME

Climate change poses a serious threat to the survival of many species and to the well-being of people around the world.

WWF's programme has three main aims:

- to ensure that industrialised nations make substantial reductions in their domestic emissions of carbon dioxide—the main global warming gas—by 2010

- to promote the use of clean renewable energy in the developing world

- to reduce the vulnerability of nature and economies to the impacts of climate change.

WWF's mission is to stop the degradation of the planet's natural environment and to build a future in which humans can live in harmony with nature, by:

- conserving the world's biological diversity
- ensuring that the use of renewable resources is sustainable
- promoting the reduction of pollution and wasteful consumption.

Is Wildlife Affected
by Climate Change?

Scientific reports on the effects of climate change are frequently cited because of their global significance. A recent study published in the journal *Nature* stated that if global warming emissions are not reduced, we stand to lose more than 1 million different species to extinction by 2050.[1] The Intergovernmental Panel on Climate Change (IPCC) issues technical reports that reflect the current scientific thinking of the world's experts.

A technical report that the IPCC issued in 2002, called *Climate Change and Biodiversity*, addresses the effects of climate change on the plants and animals of the Earth. Quite clearly, the report states: "Climate change is projected to affect all aspect of biodiversity" and "Many species have shown changes in morphology, physiology, and behavior associated with changes in climatic variables."[2] It also states that "Changes in species distribution linked to changes in climatic factors have been observed."[3]

The following excerpt from Stephen Schneider and Terry Root's 2002 book *Wildlife Responses to Climate Change* discusses the significant changes scientists have observed in the global climate. Over the past 40 years, birds have shifted ranges. Reptiles and amphibians seem to be more susceptible to climate change because they are cold-blooded and cannot move as easily or as far as birds can. Studies also report observed animal changes related to climate change—from the disappearance of tropical frogs to how the changing climate affects the body size of various mammals.

The point is that biologists are keeping records, observing patterns, and contributing to the body of knowledge on which responsible land use planning and global policy can be based.

—The Editor

1. The Nature Conservancy. *Climate Change.* 2004. Available online at http://nature.org/magazine/summer2004/climate_change/index.html.

2. Intergovernmental Panel on Climate Change (IPCC). *Climate Change and Biodiversity: Technical Paper V.* April 2002, p. 12. Available online at http://www.ipcc.ch/pub/tpbiodiv.pdf.

3. Ibid., p. 13.

Wildlife Responses to Climate Change
by Stephen Schneider and Terry Root

PREDICTING VEGETATION RESPONSES TO CLIMATE CHANGE
The Holdridge (1967) life-zone classification assigns biomes (for example, tundra, grassland, desert, or tropical moist forest) according to two measurable variables: temperature and precipitation. Other more complicated large-scale formulas have been developed to predict vegetation patterns from a combination of large-scale predictors (for example, temperature, soil moisture, or solar radiation); the vegetation modeled includes individual species, limited groups of vegetation types, or biomes. These kinds of models predict vegetation patterns that represent the gross features of actual vegetation patterns, which is an incentive to use them to predict vegetation change with changing climate, but they have some serious drawbacks as well. That is, they are typically static, not time-evolving dynamic simulations, and thus cannot capture the transient sequence of changes that would take place in reality. In addition, such static biome models occasionally make "commission errors"—they predict vegetation types to occur in certain zones where climate would indeed permit such vegetation, but other factors like soils, topography, or disturbances like fire actually preclude it. Furthermore, local patterns may influence vegetation dynamics at scales not captured in some simulations, and seed germination and dispersal mechanisms are also either not explicitly simulated or simulated only crudely with such models. Remarkably they are still able to produce generalized maps of vegetation types that do indeed resemble current or even paleoclimatic patterns in a broad sense. Their details, however, do not provide confident projections for future vegetation states. Fortunately, progress is being made to include some of the deficiencies mentioned above, and so-called dynamical global vegetation models are being developed to treat the transient nature of vegetation change that would likely accompany climate change.

PREDICTING ANIMAL RESPONSES TO CLIMATE CHANGE

Scientists of the U.S. Geological Survey (USGS), in cooperation with Canadian scientists, conduct the annual North American Breeding Bird Survey, which provides distribution and abundance information for birds across the United States and Canada. From these data, collected by volunteers under strict guidance from the USGS, shifts in bird ranges and densities can be examined. Because these censuses were begun in the 1960s, these data can provide a wealth of baseline information. [J.] Price (1995) has used these data to examine the birds that breed in the Great Plains. By using the present-day ranges and abundances for each of the species, Price derived large-scale, empirical-statistical models based on various climate variables (for example, maximum temperature in the hottest month and total precipitation in the wettest month) that provided estimates of the current bird ranges. Then, by using a GCM [global climate model] to forecast how doubling of CO_2 would affect the climate variables in the models, he applied statistical models to predict the possible shape and location of the birds' ranges.

Significant changes were found for nearly all birds examined. The ranges of most species moved north, up mountain slopes, or both. The empirical models assume that these species are capable of moving into these new areas, provided habitat is available and no major barriers exist. Such shifting of ranges and abundances could cause local extirpations in the more southern portions of the birds' ranges, and, if movement to the north is impossible, extinctions of entire species could occur. We must bear in mind, however, that this empirical-statistical technique, which associates large-scale patterns of bird ranges with large-scale patterns of climate, does not explicitly represent the physical and biological mechanisms that could lead to changes in birds' ranges. Therefore, such detailed maps should be viewed only as illustrative of the potential for very significant shifts with doubled CO_2 climate change scenarios. More refined techniques that also attempt to include actual mechanisms for ecological changes are discussed later.

Reptiles and amphibians, which together are called herpeto-fauna (or herps), are different from birds in many ways that are important to our discussion. First, because herps are ectotherms—meaning their body temperatures adjust to the ambient temperature and radiation of the environment—they must avoid environments where temperatures are too cold or too hot. Second, amphibians must live near water, not only because the reproductive part of their life cycle is dependent on water, but also because they must keep their skin moist to allow them to respire through their skin. Third, herps are not able to disperse as easily as birds, and the habitat through which they crawl must not be too dry or otherwise impassible (for example, high mountains or busy superhighways).

As the climate changes, the character of extreme weather events, such as cold snaps and droughts, will also change, necessitating relatively rapid habitat changes for most animals. Rapid movements by birds are possible because they can fly, but for herps such movements are much more difficult. For example, R. L. Burke (then [1995] at University of Michigan, Ann Arbor) noted that during the 1988 drought in Michigan, many more turtles than usual were found dead on the roads. He assumed they were trying to move from their usual water holes to others that had not yet dried up or that were cooler. For such species, moving across roads usually means high mortality. In the long term, most birds can readily colonize new habitat as climatic regimes shift, but herp dispersal (colonization) rates are slow. Indeed, some reptile and amphibian species may still be expanding their ranges north even now, thousands of years after the last glacial retreat.

R. L. Burke and T. L. Root performed a preliminary analysis of North American herp ranges in an attempt to determine which, if any, are associated with climatic factors such as temperature, vegetation-greening duration, and solar radiation. Their evidence suggests that northern boundaries of some species ranges are associated with these factors, implying that climate change could have a dramatic impact on the occurrence of these species. Furthermore, most North American turtles and several other reptile species could exhibit vulnerability to climate change because the temperature experienced as they develop inside the egg determines their sex. Such temperature-dependent sex determination makes these animals uniquely sensitive to temperature change, meaning that climate change could potentially cause severely skewed sex ratios, which could result in dramatic range contractions. Many more extinctions are possible in herps than in birds because the forecasted human-induced climatic changes could occur rapidly when compared with the rate of natural climatic changes, and because the dispersal ability of most herps is painfully slow, even without considering the additional difficulties associated with human land-use changes disturbing their migration paths.

In general, animals most likely to be affected earliest by climate change are those in which populations are fairly small and limited to isolated habitat islands. There are estimates that a number of small mammals living near isolated mountain-tops (which are essentially habitat islands) in the Great Basin would become extinct given typical global change scenarios. Recent studies of small mammals in Yellowstone National Park show that statistically significant changes in both abundances and physical sizes of some species occurred with historical climate variations (which were much smaller than most projected climate changes for the next century), but there appear to have been no simultaneous genetic changes. Therefore, climate change in the 21st century could likely cause substantial alteration to biotic communities, even in protected habitats such as Yellowstone National Park.

CURRENT ANIMAL RESPONSES TO CLIMATE CHANGE

Animals are showing many different types of changes related to climate. These include changes in ranges; abundances; phenology (timing of an event); morphology and physiology; and community composition, biotic interactions, and behavior. Changes are being seen in all different types of taxa, from insects to mammals, and on many of the continents. For example, the ranges of butterflies in Europe and North

America have been found to shift poleward and upward in elevation as temperatures have increased. From 1979 to 1989, population densities of the Puerto Rican coqui (*Eleutherodactylus coqui*) showed a negative correlation with the longest dry period during the previous year. Similarly, the disappearance of the golden toad (*Bufo periglenes*) and the harlequin frog (*Atelopus varius*) from Costa Rica's Monteverde Cloud Forest Reserve seemed to be linked to the extremely dry weather associated with the 1986–87 El Niño–Southern Oscillation. Birds' ranges reportedly have extended poleward in Antarctica, and Europe. For instance, the northern movement of the spring range of barnacle geese (*Branta leucopsis*) along the Norwegian coast correlates significantly with an increase in the number of April and May days with temperatures above 6°C [43°F]. Reproductive success of the California quail (*Calipepla californica*) is positively correlated with the previous winter's precipitation. Rainfall affects the chemistry of plants eaten by quail, with the plants producing phytoestrogens, compounds similar to hormones that regulate reproduction in birds and mammals. Drought-stunted plants tend to have higher concentrations of these compounds. The northern extension of the porcupine's (*Erethizon dorsatum*) range in central Canada has been associated with a warming-associated poleward shift in the location of tree line. In the United Kingdom, the dormouse (*Muscardinus avellanarius*) has disappeared from approximately half of its range over the last 100 years. This disappearance appears to be linked to a complex set of factors including climatic changes, fragmentation, and the deterioration and loss of specialized habitat.

Warmer conditions during autumn–spring adversely affect the phenology of some cold-hardy insects. Experimental work on spittlebugs (*Philaenus spumarius*) found that they hatched earlier in winter-warmed (3°C [37.4°F] above ambient) grassland plots. Chorusing behavior in frogs, an indication of breeding activities, appears to be triggered by rain and temperature. Two frog species, at their northern range limit in the United Kingdom, spawned 2 to 3 weeks earlier in 1994 than in 1978.

Three newt species also showed highly significant trends toward earlier breeding, with the first individuals arriving 5 to 7 weeks earlier over the course of the same study period. This study also examined temperature data, finding strong correlations with average minimum temperature in March and April (negative) and maximum temperature in March (positive) for the two frogs with significant trends, and a strong negative correlation between lateness of pond arrival and average maximum temperature in the month before arrival for the newts. Using less precise methods, a family of naturalists in England recorded the timing of first frog and toad croaks for the period from 1736 to 1947. The date of spring calling for these amphibians became earlier over time, and was positively correlated with spring temperature, which was positively correlated with year. Changes in phenology or links between phenology and climate have been noted for earlier breeding of some birds in the United Kingdom, Germany and the United States. Changes in bird migration have also been noted with earlier arrival dates of spring migrants in the United States, later autumn departure dates in Europe, and changes in migratory patterns in Africa.

The effect of temperature on the metabolism of dormant horned toads in Brazil was found to be stronger than the effect on resting toads at most temperatures. Reptile physiology is temperature sensitive also. Painted turtles grew larger in warmer years, and during warm sets of years turtles reached sexual maturity faster. Physiological effects of temperature can also occur while reptiles are still within their eggs. Leopard geckos (*Eublepharis macularius*) produced from eggs incubated at a high temperature of 32°C [89.6°F] showed reproductive behavioral changes and possible female sterility. Spring and summer temperatures have been linked to variations in the size of the eggs of the pied flycatcher (*Ficedula hypoleuca*). The early summer mean temperatures explaining approximately 34% of the annual variation in egg size between the years 1975 and 1994. Body mass, which correlates with many life-history traits including reproduction, diet, and size of home ranges of the

North American wood rat (*Neotoma* spp.) has shown a significant decline inversely correlated with a significant increase in temperature over the last 8 years in one arid region of North America. In studies of spring temperature effects on red deer (*Cervus elaphus*) in Scotland, juvenile deer grew faster in warm springs leading to increases in adult body size, a trait positively correlated with adult reproductive success. In Norway, red deer born following warm winters (that have more snow) were smaller than those born after cold winters—a difference persisting into adulthood.

Differential responses by species could cause existing animal communities to undergo a reformulation. Peach-potato aphids grown on plants kept in elevated CO_2 (700 ppm [parts per million]) showed a reduced response to alarm pheromones in comparison to those grown on plants in ambient CO_2 (350 ppm). The aphids were more likely to remain on leaves, rather than move away, in response to the pheromones, possibly making them more susceptible to predators and parasitoids. Temperature and dissolved-oxygen concentrations can alter the behavior of amphibian larvae, and changes in thermal environments can alter the outcome of predator-prey interactions. Climate change may be causing mismatching in the timing of breeding of great tits (*Parus major*) in the United Kingdom and other species in their communities. [E.] Post et al. (1999) documented a positive correlation between gray wolf (*Canis lupus*) pack size in winter and snow depth on Isle Royale (U.S.). In years with deeper snow, wolves formed larger packs, which led to more than three times as many moose kills.

TOP-DOWN APPROACHES

The biogeographic approach just summarized is an example of a top-down technique (like that of the Holdridge life-zone classification), in which data on abundances or range limits of species, vegetation types, or biomes are overlain on data of large-scale environmental factors such as temperature or precipitation. When associations among large-scale biological and climatic patterns are revealed, biogeographic rules expressing these correlations graphically or mathematically can be used to forecast changes driven by given climate changes.

BOTTOM-UP APPROACHES

The next traditional analysis and forecasting technique is often referred to as bottom-up. Small-scale ecological studies have been undertaken at the scale of a plant or even a single leaf to understand how, for example, increased atmospheric CO_2 concentrations might directly enhance photosynthesis, net primary production, or water-use efficiency. Most studies such as these indicate increases in all these factors, increases that some researchers have extrapolated to ecosystems.

However, at the scale of a forest, the relative humidity within the canopy, which significantly influences the evapotranspiration rate, is itself regulated by the forest. In other words, if an increase in water-use efficiency decreased the transpiration from each tree, the aggregate forest effect would be to lower relative humidity. This, in turn, would increase transpiration, thereby offsetting some of the direct CO_2/water-use efficiency improvements observed experimentally at the scale of a single leaf or plant. Regardless of the extent to which this forest-scale negative feedback effect (or "emergent property" of the coupled forest atmosphere system) will offset inferences made from bottom-up studies of isolated plants, the following general conclusion emerges: the bottom-up methods may be appropriate for some processes at some scales in environmental science, but they cannot be considered to produce highly confident conclusions without some sort of validation testing at the scale of the system under study.

Are Polar Bears Affected by Climate Change?

Most of us will probably never get to see a polar bear in its natural habitat, but what is happening to them should be of concern. We have a variety of reasons to be worried. Some people will be concerned because they believe that polar bears are living creatures with a natural right to exist. Others may be concerned simply because what is threatening the polar bears is also threatening us. Global warming is just that—global.

The overall projections for global warming are changes in temperature and patterns of rainfall. As stated in the Intergovernmental Panel on Climate Change (IPCC) report *Climate Change and Biodiversity*: "Northern Hemisphere snow cover, permafrost, and sea-ice extent are projected to decrease further. The Antarctic ice sheet is likely to gain mass because of greater precipitation, while the Greenland ice sheet is likely to lose mass. . . ."[1]

As global temperatures increase, changes are being observed in the Greenland ice sheet. Polar bears, the largest terrestrial (land) carnivores in the world, spend a lot of time on ice packs in the Northern Hemisphere. The following article discusses the negative effect that ice melting too early in the season has on mother bears in dens with their cubs. Without the ice and the seals on the ice, the bears have less food. Biologists have observed that over the past 20 years, polar bears have been losing weight. They have also found a correlation between the bears' weight loss and the time of year that the ice packs melt.

As in all ecosystems, when there is a problem observed with the top predators, like wolves, grizzlies, or orca whales, it signals other problems in the system. Such studies are important because they add to the scientific data used to make policy on how to address global change.

—The Editor

1. Intergovernmental Panel on Climate Change (IPCC). *Climate Change and Biodiversity: Technical Paper V.* April 2002, p. 11. Available online at *http://www.ipcc.ch/pub/tpbiodiv.pdf.*

The Incredible Shrinking Polar Bears
by Jim Morrison

In Canada's Hudson Bay, a long-term study confirms they are losing weight and bearing fewer cubs as global warming melts away their icy habitat. Is this a preview of what other populations of polar bears will soon be facing?

By March when a mother polar bear departs her earthen den with nursing cubs into subzero temperatures along Canada's Hudson Bay, she has fasted for eight months and lost more than half her weight, as much as 400 pounds [181 kg]. After a week or two of acclimation around the den, she heads to the frozen water. Fortunately, her emergence onto the sea ice along the continental shelf coincides with the birth of ringed seals, her primary prey.

From April until summer when the ice breaks up in the bay and the seals disappear into open water, she is in a desperate race against time to pack on enough pounds of fat to get her through the long summer and fall fast. If she becomes too lean, she'll stop producing milk and her cubs will die.

Hunting is hard work. With her massive limbs and wide paws, evolved for prowling the ice and swimming, she expends twice the energy to walk than most other mammals. So she stalks her prey slowly, relying on her extraordinary sense of smell—she can sniff seals in their subnivean [beneath the snow] lairs up to a mile away. Often, she will remain motionless on her stomach for long stretches beside a breathing hole, waiting for a seal to surface.

Polar bears, the largest terrestrial carnivores in the world, abandoned the land for this harsh life between 200,000 and 300,000 years ago. Taxonomists have since named them *Ursus maritimus*, the "bear of the sea." The creatures have been known to swim as far as 60 miles [97 km] in a day. But they're really the bear of the ice. The shifting floes of the Arctic region are home to these massive animals. In areas of Hudson Bay, however, the ice is now abandoning them, threatening the long-term survival of the polar bears that have survived for eons in the region. Their plight is a preview of the challenges other polar bear populations that live farther north will face in the coming decades.

Global warming, experts believe, is causing the ice pack to melt an average of two weeks earlier each July than 20 years ago. In one study, Josefino Comiso, a researcher at NASA's Goddard Space Flight Center in Maryland, used satellite data to track trends in minimum sea ice cover and temperature over the Arctic from 1978 to 2000. Last year, he reported that the perennial sea ice is melting faster than previously thought due to rising temperatures and interactions between ice, ocean and the atmosphere that accelerate the melting process.

The timing of the early ice melt couldn't be worse for polar bears, especially pregnant and nursing females and their cubs. In Hudson Bay, fat-rich ringed seal pups wean in May and then emerge from their lairs. A month later, adult seals haul out on the ice to molt, offering bears a second course. But when the ice disappears early, so does the bears' larded buffet.

"Nobody really knows what percentage of the fat a bear stores and uses through the year is captured in the spring," says Ian Stirling, a Canadian Wildlife Service scientist who has studied polar bears for more than 30 years. "It's certainly well over 50 percent and may be as high as 70 or 75 percent. All we know is it's terribly important."

Stirling and his colleagues also know that polar bears along the Hudson Bay are lighter and in poorer condition than they were 20 years ago. They know because their study of the bears—looking at a single population for more than two decades—is unique in its longevity. For each week the ice breaks up earlier, the researchers have found, the bears come ashore 22 pounds [10 kg] lighter. And if climate models created in recent years by several scientists in the United States and Great Britain prove to be correct, the area will be three to five degrees warmer within 50 years. With every degree increase, the ice breakup will occur one week earlier.

The 1,200 bears of the western Hudson Bay live at one of the southernmost extremes of the species' range. As the planet continues to warm, what happens there will no doubt happen farther north to the rest of the estimated 25,000 bears that live in 20 distinct circumpolar populations. Already, ice conditions in the Beaufort Sea off the north coast of Alaska, home to another major polar bear population, are changing dramatically.

The longer summers in the Hudson Bay area are particularly challenging for pregnant and nursing females and their cubs. Mothers weigh between 300 and 350 pounds [136 and 159 kg] after leaving the den and need to gain between 100 and 200 pounds [45 and 91 kg] before the thaw arrives. Healthy pregnant females often gain more than 400 pounds [181 kg] of fat alone. So a mother may spend half her time hunting, catching a seal every few days. She can gorge on more than 60 pounds [27 kg] of seal meat at a single sitting. (A typical 1,200-pound [544-kg] male bear can down 150 pounds [68 kg] at a time.)

In most areas, polar bear cubs are weaned at about two and a half years of age. In some years, fewer than half of those cubs may live to become adults. The main cause of death is a lack of food. When mothers become too lean, they simply stop nursing and abandon their cubs. "Lighter cubs have lower survival rates and light cubs come from leaner pregnant females," says Andrew Derocher, a University of Alberta scientist who has studied bears for 20 years. "For a polar bear, fat is where it's at."

"If all the ice goes in Hudson Bay, as the forecasts predict, then there won't be any polar bears there," Stirling says. "And there's no place they can go. People often ask, 'Can't they just go farther north?' But the answer is no. That habitat is already occupied by other polar bears." While some bears do cross the North Pole, the conditions for survival there are so extreme that the region is inhospitable even for them. "For the long term, I think the situation looks very bad for polar bears," he adds.

The connection between global warming and the bears' deteriorating condition could be proved only because of Stirling's desire to build a unique database about the lives of polar bears—and because of geographic luck. The Hudson Bay bears are the most studied population in the world, thanks to their accessibility and Stirling's remarkable ability to acquire funding year after year. "The Hudson Bay database exists because of the determination of a single individual," Derocher says. "Ian's study has provided us with insights not possible from other polar bear populations."

Other populations move across large expanses of moving ice. But along the bay, the entire population comes ashore in one concentrated area for a few months annually. That makes conducting research there efficient—a necessity for cash-strapped researchers. Over more than two decades, Stirling and his associates have captured 80 percent of the adult bears in the area. Each has been tagged, measured, weighed, checked for fat.

The luck came in the location of the research. As it happens, the western area of Hudson Bay has been a hot spot—more affected by global warming than nearby regions so the changes there have been more dramatic. While statistics show that the world generally is getting warmer, temperature changes vary by location. Southeastern Hudson Bay, for example, actually became cooler over the last few decades, though recently the area has been warming as well.

In the early 1980s, Stirling began noticing that bears seemed lighter. Cubs were taking longer to wean so females were reproducing less frequently. There was no Eureka moment, he says, just a gradual realization that a trend seemed to be developing.

He got another clue after Mount Pinatubo in the Philippines blew its top in 1991 and the particulate matter it sent into the atmosphere cooled the planet. The following summer the ice on Hudson Bay melted almost a month later, extending the bears' hunting season in the process. The effect was dramatic. The bears were heavier, had more cubs and more of the youngsters survived those brutal first years. But that was still only a piece of a tricky, subtle puzzle.

"It took 20 years to have enough data to determine that this was happening," Stirling says, "because there always is quite a bit of annual fluctuation. You can't detect an underlying trend such as this unless you've been able to look at it for a long time."

Because polar bears are at the top of the food chain, their plight provides a strong, early warning about the stress climate change is putting on the entire Arctic region. "Usually when you take large predators out of an ecosystem you get major changes. But exactly what those would be I don't think anybody would say," says Stirling. "Clearly, there would be huge changes in the Arctic ecosystem without polar bears."

Stirling is now nearing retirement age and says it's time for some of the younger researchers to step up to the plate. In Derocher and Nick Lunn, his colleague at the Canadian Wildlife Service, he has two eager disciples determined to carry on his work. "There's no other data like this on any Arctic species. If you let it go, even for a year, it's gone. You can't fill that gap," says Lunn, who wants to spend more time studying the reproduction of individual animals over their lifespans. Initially, he notes, it may be that not every bear will be affected in the same way by the warming trend.

Derocher, meanwhile, plans to further study the effects of climate change on bear populations. "I believe," he says, "that a better understanding of interactions between sea ice, polar bears and their prey are the key issues for understanding how the bears will respond to climate change over the coming years." He also hopes to continue investigating the effects of toxic chemicals on the bears and other Arctic wildlife.

These days, Derocher spends a lot of time thinking about what it's like to be a polar bear, walking on the sea ice, using smell to find dinner during the howling Arctic night when the temperature is minus 40 degrees F [-40°C]. "You can't help but respect an animal that makes a living in this kind of environment," he says. "It's a scenario that humans don't relate to very well."

WORLD CLASS NAVIGATORS

When it comes to establishing home ranges, polar bears are the undisputed champions of the bear world. While a grizzly's territory in the Rocky Mountains might encompass as much as 1,000 square miles [2,590 km^2], in some cases a polar bear's range in the Arctic might extend across 100,000 or more square miles [258,999 or more km²], depending upon the availability of ice and food. "They are good navigators," says Canadian Wildlife Service scientist Ian Stirling, "although we do not yet know how they do it." He and other bear biologists do know that the animals have remarkable capabilities. Using satellite technology, researchers tracked one female polar bear for four months as she traveled from the Beaufort Sea coast in Alaska to northern Greenland, passing within 150 miles [241 km] of the North Pole along the way, a distance of about 3,000 miles [4,828 km].

What Impact Will Climate Change Have on Fish Populations?

A 2002 report issued by the Intergovernmental Panel on Climate Change (IPCC) concluded that, "Climate changes will have a pronounced effect on freshwater ecosystems through alterations of the hydrological processes."[1] It also stated that "Increased temperatures will alter thermal cycles of lakes . . . and thus affect ecosystem structure and function."[2] These predictions from the IPCC, a group of more than 2,000 experts in various environmental fields, serve as scientific validation for what some biologists and fisheries managers have been noticing for years.

Part of science is observing patterns, and research scientists have been collecting data from a variety of sources to gain information about the effect of climate change on fisheries. The following article discusses the impact that global warming is having on fish populations in North America. Using such diverse sources as 150-year-old records from trappers and computer-generated models of weather systems, researchers have seen some patterns that are of concern. They also make some predictions about what may happen if the patterns, or trends, they observe are allowed to continue.

Scientists predict that increases in water temperatures could cause changes in the distribution of fish and the way that lake waters stratify, or separate into layers of different temperatures. Lake stratification affects the oxygen content of the water and the areas in which fish can find places to get away from winter ice. Changes in weather patterns will impact fish breeding areas, food supplies, and the presence of predators. It is not just the changes in temperature, but the actual time at which the temperatures change and the ice melts that affects fish and their food sources.

People who manage fisheries discussed the concerns at a 2003 meeting of the American Fisheries Society, comparing notes on the changes they have observed in the fisheries around the country. "The silver lining," says Robert Wood of the University of Maryland, "is that we now recognize that climate change can have profound effects on fisheries."

As researchers continue to collect data and observe patterns of change, perhaps fishery managers can prepare their industry for the changes as politicians and policymakers work to minimize them.

—The Editor

1. Intergovernmental Panel on Climate Change (IPCC). *Climate Change and Biodiversity: Technical Paper V*. April 2002. Available online at *http://www.ipcc.ch/pub/tpbiodiv.pdf*.

2. Ibid.

North American Fish Feel the Heat
by T. Edward Nickens

How will climate change affect fish? New studies from across the continent provide some dramatic and troublesome answers.

John Magnuson doesn't have to fire up the latest computer modeling program to see how global warming is affecting his world—he can look out his office window at the University of Wisconsin–Madison and see the broad, blue sweep of Lake Mendota. Magnuson is a limnologist, a scientist specializing in the ecology of lakes and ponds, and since 1853, he reports, there has been a 25 percent decrease in the amount of time the lake remains frozen over during the winter.

That's tough on ice fishermen, who flock to Mendota. But it might be even tougher on Lake Mendota's fish. Magnuson is one of a growing number of scientists who are concerned that global warming is setting the stage for large-scale impacts on fish populations, such as Lake Mendota's perch, walleye and lake herring. These researchers are turning to data as varied as 150-year-old fur-trapping records and computer modeling of hemispheric-scale weather systems to forecast how fish might respond to global climate change. Their findings reveal a spectrum of potential changes in fish populations that should

concern—if not astound—anglers, conservationists, commercial fishermen and others who care about the future of fish in
North America.

Unlike the visible and often immediate impacts of pollution or urban sprawl, global warming's effects on fish ecology
and habitat are likely long-term, complex and, in some cases,
subtle. But they could prove just as dramatic. Scientists believe
that rising water temperatures could shift fish distributions
around the country, moving the boundary that separates cold-
water species (such as trout and walleye) from warm-water
fish (such as largemouth bass and even subtropical tilapia)
north by as much as 300 miles [483 km] by mid-century.

As the Earth's climate warms and large-scale atmospheric
circulation patterns change, a network of ecological changes
might follow that would impact every stage of fish biology.
New patterns of seasonal flooding could scour fish eggs from
southern Appalachian trout waters and Pacific Northwest
salmon streams. Changing ice regimes could impact whitefish
on a population scale. Deep lakes might become increasingly
depleted of oxygen. In the Chesapeake Bay, larval striped bass
and perch could suffer from waters that offer less food because
of earlier spring temperatures. And native species of fish across
the continent could face greater competition from invaders
from more southerly climes.

"These findings are based on forecasts, modeling and
possible scenarios," cautions Frank Rahel of the University of
Wyoming. "We're talking about potential effects, and there are
major unknowns." Although the specifics of how fish will
respond to global warming are still in question, few researchers
doubt that fish face a changing thermal world. A consensus of
concern formed at last summer's meeting of the American
Fisheries Society, during a special symposium where
researchers compared notes about a changing climate's
impacts on fisheries. "What we're seeing," says Magnuson,
"is that researchers from all over the country—the Gulf,
Far North, East, West, North Atlantic, North Pacific, the
Chesapeake—now have stories to tell about how climate

change is impacting fish or will in the future. I thought we might be living in a little microcosm of change up here on Lake Mendota. But no part of North America escapes."

The potential impacts range from primary, temperature-related effects to insidious secondary and tertiary possibilities. Consider water temperature. As air temperatures warm, many stream, river and lake temperatures will rise accordingly. For the trout of the Rocky Mountain west, Rahel found, the results could be dramatic. Rahel and his graduate student Christopher Keleher calculated that a 1-degree C [33.8°F] increase in the average July air temperature of the region would reduce thermally suitable trout habitat in the Rockies by 17 percent. A 2-degree C [35.6°F] increase would wipe out more than a third. Warm the average July climate by 3 degrees C [37.4°F]— well within the range of many meteorological models—and half of the Rockies' suitable trout range evaporates before today's first-graders start collecting Social Security.

"These are very substantial losses for cold-water species," Rahel admits. "We were surprised at the high numbers." When the researchers produced maps showing the future potential distribution of cold-water fish given a 3-degree C increase in average July air temperature, the document looked like a piece of tattered lace.

And the nature of the remaining habitat is as much of a concern as the loss of habitat. As native cold-water fish are pushed farther and farther up the sides of the mountains, populations will become increasingly fragmented and isolated. Already conservationists are concerned about imperiled Rocky Mountain trout populations such as the Yellowstone cutthroat, Bonneville cutthroat and federally endangered greenback cutthroat trout. With warming water temperatures, Rahel explains, not only is there a loss of suitable habitat, but "tributaries are left dangling out in the landscape, isolated from other populations that would normally provide a mix of genetic material. And once you have a series of small, isolated populations, they become vulnerable to extinction like lights blinking out on a Christmas tree."

It's easy to understand that waters might get too warm for cold-water fish. Other potential impacts of a warming climate are far less obvious. Across the country from those Rocky Mountain streams is the Chesapeake Bay, one of the largest and most productive estuaries in the world. Each spring a large pulse of fresh water flows from the bay's tributaries, bringing a flush of nutrients that kicks off a series of plankton blooms—first a bloom of minute floating plants, or phytoplankton, and then a flush of tiny animal organisms, or zooplankton, that feed on the microscopic plants. Millions of hungry fish larvae—striped bass, white perch, menhaden, spot and others—depend on certain kinds of zooplankton, but not at the same time, says Robert Wood of the University of Maryland. Spot and Atlantic menhaden, for example, spawn far out on the coastal Atlantic shelf during winter. Their eggs are transported to the bay by wind-driven currents and arrive as hungry larvae at their riverine nursery grounds in March and early April. Striped bass and white perch spawn in April, and don't develop into zooplankton-eating larvae until May and June.

But what if winters get shorter and spring comes earlier, as climate warming models suggest? "That would favor coastal spawning species such as spot and menhaden over striped bass and perch," Wood explains, although he points out that "scientists can't be sure which dominoes will fall, and when. And it's not simply a matter of what happens in the spring, but a matter of the timing of these processes."

And how they relate to other impacts of global warming. Wood lists a series of changes that could befall the bay. Paired to global warming is sea level rise, which will consume the tidal wetlands that serve as nursery areas for fish as varied as mummichog and spotted seatrout. Warming water temperatures could push out cool-water species such as soft clams and striped bass, even though some fisheries might actually benefit from a warmer Chesapeake Bay. Brown and pink shrimp could increase. Blue crabs tend to fare better during warmer winters. "But we don't know enough to predict the particular response of the ecosystem to global climate change in the Chesapeake Bay," says Wood. "It's like a spider's web. Break some of the connections and we're not sure which parts of the ecosystem will degrade. But it's clear that cascading effects could transform the ecosystem in ways that we can't foresee right now."

Changes in some ecosystems already are visible, if you look as hard—and as far—as Wisconsin's John Magnuson has. In 2000, Magnuson led an international research team that uncovered evidence that lakes and rivers in the Northern Hemisphere are freezing an average of nearly nine days later, and thawing ten days earlier, than they did a century and a half ago. This conclusion came from studying records from around the globe, including Hudson's Bay Company shipping logs and freeze dates from ancient Japanese religious sites. In one intriguing case, a Madonna figure had been carried across a lake on the Germany-Switzerland border each year that the lake froze as far back as the ninth century, giving savvy researchers a data set of total ice cover.

"Ice is very responsive to changes in climate, and so are fish," explains Magnuson. But even more importantly, ice records provide a longer record than do conventional fishery statistics. Trends that might not appear in studies that examine year-to-year or even decade-to-decade data can show up when the time scale is broadened to a century or more. "Fishermen see behavioral changes in fish over their lifetimes, so they know that fish are very responsive to climate change and variability," says Magnuson. "But the biggest changes are the long-term shifts, even though they are more difficult to observe."

A changing regime of ice cover of lakes and rivers could affect fish through hidden, second-order impacts in the same way the warming of the Chesapeake Bay might skew fish ecology. In Lake Michigan's Grand Traverse Bay, Magnuson says, whitefish have had poor hatching success in years when there is relatively little ice cover. "No ice, and winter winds roil the water, mixing it up," he explains. "When the water surface is frozen, the eggs sit safely on the bottom in a very still environment."

Another potential impact would be a shift in how lake waters stratify. During the summer, warmer water layers the tops of lakes, while cooler water settles towards the bottom. Those cooler waters provide a thermal refuge for many cold-water fish, such as lake trout, whitefish and salmon. If the lakes are covered with ice for shorter periods of time, Magnuson says, the growing season lengthens and isolated deep waters run the risk of running out of oxygen. "The trout, whitefish and salmon that need those cool-water refuges will suffer."

These effects, researchers say, signal a paradigm shift for fisheries managers. "We've always assumed that human fishing activity was the major regulator of fisheries abundance," explains Magnuson. But these new studies show that climate change and variability affect fish populations in significant ways. How humans manage fish will need to change. In the salmon fisheries of the Pacific Northwest and Canada, harvest rates for certain stocks of salmon have commonly been as high as 60 to 80 percent of the total population. "We've been managing salmon fisheries for 50 years by making decisions based on employing people and producing seafood," explains Richard Beamish, with Fisheries and Oceans Canada. "People have assumed that fishing harvest rates were so high, and the ocean so vast, that the ocean environment wasn't even an issue." Now researchers like Beamish contend that the effects of climate on the ocean are as important as fisheries impacts.

Beamish recently analyzed pink, chum and sockeye salmon harvest rates to see how fish populations respond to broad shifts in Pacific weather patterns. Plotting surface water temperatures, atmospheric circulation of winter winds and changes in the Aleutian Low (the north Pacific's major winter weather system), Beamish learned that Pacific salmon stocks responded dramatically to four major shifts in these weather indices—in 1947, 1977, 1989 and 1998. In certain instances these "regime shifts," as Beamish calls them, prompted increases in salmon numbers. In others, fish populations crashed. In 1985 and 1986, the highest total catches in the history of the Canadian salmon fishery were tallied, about

105,600 tons for each of those years. Following the last major climatic shift in the region, in 1998, the total catch was a meager 25,500 tons. Last year it dropped even lower, to 7,500 tons, the lowest in history.

The mechanics of how salmon respond to such changes isn't yet clearly understood. But recognizing that salmon abundance fluctuates according to the dynamics of the Pacific Ocean's weather patterns, Beamish says, is a critical first step to bringing scientists and fisheries managers together. "Now we have to learn how to incorporate climate change effects into our strategies for managing fish populations," Beamish says.

It's a welcome sea change. For years, scientists concerned about the impacts of climate on fish populations felt like voices crying in far-flung wildernesses. No longer. And if the thought of casting to smallmouth bass in a river that once held Wyoming cutthroat trout is troublesome, at least the acknoledgment of the issue will empower fisheries managers. "The silver lining," says Robert Wood, "is that we now recognize that climate change can have profound effects on fisheries."

Will Climate Change Impact Our Oceans?

In the past, the United States might have felt protected from the troubles of the most of the world because the nation is surrounded on either side by huge oceans. But today, for many reasons, the United States is no longer truly isolated from global problems. And now the oceans themselves are something about which we must be concerned. In addition to overfishing and pollution, global warming has been added to the list of threats to the oceans.

The following selection is from a 2003 report by the Pew Oceans Commission, *America's Living Oceans: Charting a Course for Sea Change*. It identifies the potential effects of global warming on the oceans and coastal areas. Coral reefs are sensitive ecosystems that are already experiencing, among other problems, bleaching and susceptibility to diseases from increased water temperatures. Global warming is also predicted to alter weather patterns, which will cause changes in global wind and water current patterns. Temperature shifts will also affect the distribution of species and their reproductive cycles. Rising sea levels and new precipitation patterns will flood coastal wetlands or leave them without enough water. Although it is difficult to predict what increased atmospheric carbon will do to the oceans, some projections show chemical changes that would affect the growth of organisms that need calcium to survive.

The significance of these predictions and scientific concerns is that our human activities on land, such as the burning of fossil fuels, will seriously impact the oceans. Oceans cover almost three-quarters of the Earth and, as one might imagine, they play a huge role in maintaining the health of the planet. Intergovernmental Panel on Climate Control (IPCC) studies and Pew Oceans Commission reports add the scientific opinions of experts to policymakers' field of facts. Decisions made at all levels—local, national, and international, and even individual—impact the issues of global warming.

—The Editor

1. Intergovernmental Panel on Climate Change (IPCC). *Climate Change and Biodiversity: Technical Paper V*. April 2002, p. 13. Available online at http://www.ipcc.ch/pub/tpbiodiv.pdf.

America's Living Oceans:
Charting a Course for Sea Change

from the Pew Oceans Commission

Global air temperature is expected to warm by 2.5 to 10.4°F (1.4 to 5.8°C) over the 21st century, affecting sea-surface temperatures and raising the global sea level by 4 to 35 inches (9 to 88 cm). Such climate change will create novel challenges for coastal and marine ecosystems already stressed by overfishing, coastal development, and pollution.

Based on observations, scientists expect that this rapid climate change will result in the extinction of some species and serious, if not catastrophic, damage to some ecosystems. Important coastal and ocean habitats, including coral reefs, coastal wetlands, estuaries, and mangrove forests will be particularly vulnerable to the effects of climate change. These systems are essential nurseries for commercial fisheries and support tourism and recreation. Wild fisheries and aquaculture will be affected as well. Climate change will modify the flow of energy and cycling of materials within ecosystems—in some cases, altering their ability to provide the ecosystem services we depend upon.

We know that climate change is no stranger to Earth. Since life began, ice ages and hot spells have affected the distribution of organisms as well as their interactions. However, today human activities that increase the emission of greenhouse gases, such as carbon dioxide, methane, and nitrous oxide, are spurring changes with a rapidity rarely experienced in Earth's history. Such high rates of change bring with them great unpredictability.

In August 2002, The Pew Center on Global Climate Change completed a report entitled *Coastal and Marine Ecosystems and Global Climate Change: Potential Effects on U.S. Resources.* It identifies the critical implications of climate change on the coastal zone and open ocean.

The authors of this report drew a number of conclusions, which we summarize below.

CORAL REEFS ARE AT PARTICULAR RISK
FROM GLOBAL CLIMATE CHANGE

Recent episodes of bleaching and high mortality of coral animals have been linked to higher temperatures. Although coral reefs are capable of recovery from bleaching events, prolonged or repeated bleaching can lead to mortality. Recent estimates suggest an increase in mean sea-surface temperature of only 2°F (1°C) could cause the global destruction of coral reef ecosystems.

Sea-level rise also poses a potential threat to coral reefs, which need the light that penetrates relatively shallow water. The problem of sea-level rise is likely to be made worse by the effects of increased atmospheric CO_2 on marine chemistry. A doubling of atmospheric CO_2, for example, could reduce coral-reef calcification (i.e., growth) by 20 to 30 percent. Although in the past, corals have been able to build their reef masses upward to keep up with rising sea levels, such slow-downs in growth induced by climate change could result in many reefs losing this race.

Increased coastal erosion associated with sea-level rise could also degrade water quality near coral reefs by increasing turbidity and sedimentation. Many coral reefs are also vulnerable to other human and natural stressors, such as coastal development, overfishing, pollution, and marine disease.

GLOBAL CLIMATE CHANGE IS PREDICTED TO AFFECT
PRECIPITATION, WIND PATTERNS, AND THE FREQUENCY
AND INTENSITY OF STORMS

These environmental variables are crucial to the structure, diversity, and function of coastal and marine ecosystems. The increase in air temperature will directly affect sea-surface temperatures and accelerate the hydrological cycle. Unequal heating and cooling of the Earth's surface drive much of the world's winds. The winds could be altered by surface warming, affecting wind-driven coastal and marine currents. Although the impact of climate change on tropical storms and hurricanes remains highly uncertain, maximum wind speeds could increase by 5 to 20 percent.

WARMING TEMPERATURES WILL INFLUENCE REPRODUCTION, GROWTH, AND METABOLISM OF MANY SPECIES IN STRESSFUL OR BENEFICIAL WAYS, DEPENDING ON THE SPECIES

In any particular region, some species could decline while others thrive. Warmer temperatures tend to enhance biological productivity, which could benefit some U.S. coastal ecosystems, at least over the short term. However, increases in temperature tend to increase the metabolic rates of organisms, leading to greater oxygen demands. At the same time, warmer water holds less oxygen than cooler water. Therefore, low oxygen conditions—which already afflict many coastal areas polluted by excess nutrients washed off the land—may worsen.

CLIMATE CHANGE HAS THE POTENTIAL TO BENEFIT AND TO HARM AQUACULTURE

Aquaculture could potentially benefit from climate change, as warmer temperatures tend to increase growth rates. Warming oceans could also allow the culturing of species in areas that are currently too cold.

However, warmer temperatures could also limit the culturing of some species. Summer mortality is often observed among cultivated Pacific oysters on the U.S. West Coast, which could be exacerbated by climate change. Warmer temperatures may increase the risk of marine disease among cultured (as well as native) species.

The implications of climate change for U.S. aquaculture will likely be heavily dependent upon the industry's ability to adapt its operations to suit the prevailing climate.

TEMPERATURE CHANGES WILL DRIVE SPECIES MIGRATION AND COULD CHANGE THE MIX OF SPECIES IN PARTICULAR REGIONS

Higher temperatures would be lethal to some species at the southern end of their range and would allow others to expand the northern end of their range, if they were sufficiently mobile. The geographic range of Pacific salmon, for example, is sensitive to changes in climatic conditions. Warm waters in the northern Pacific have historically been associated with a shift in salmon production from the coast of the Pacific Northwest to Alaska's Bering Sea. Similarly, warm-water fish species on the U.S. East Coast expanded north of Cape Cod during the 1950s in response to warmer sea-surface temperatures.

Thus, climate change in this century is likely to drive similar changes in species distributions, with some species contracting their ranges and others expanding. This would lead to different mixes of species that could affect predator-prey relationships, species competition, and food web dynamics. In addition, it could drive the proliferation of invasive species, including marine diseases.

Because many of our coastal communities depend upon marine species for their economic livelihood, redistribution will most certainly disrupt economies. However, it is impossible to predict how this will affect specific fisheries.

SEA-LEVEL RISE COULD THREATEN THE SURVIVAL OF MARSHES AND MANGROVES

As sea level rises, coastal marshes have the inherent ability to accrete (i.e., grow) vertically through the deposition of sediment carried downstream by rivers and streams. However, climate change is likely to change patterns of rainfall and runoff, which could limit sediment availability. Furthermore, human modifications of rivers and streams (e.g., dams) already limit sediment delivery in many areas, such as the wetlands of southern Louisiana. Continuation of this practice could limit the ability of wetlands to keep pace with rising sea levels.

Other human adaptations to climate change, such as the construction of seawalls to hold back the sea, could block inland migration of wetlands. Gradually, the wetlands would be inundated by rising seawater. They and their ecological services would be lost over time.

CHANGES IN PRECIPITATION COULD FLOOD COASTAL SYSTEMS OR LEAVE THEM IN DROUGHT

Changes in precipitation would affect runoff from land, and stratification of the water column, which affects oxygen concentrations in deep water. These changes also affect water circulation patterns and associated delivery of juvenile organisms to nursery areas. In concert with sea-level rise, increased runoff from land would shrink estuarine habitats, diminishing their ability to support coastal animal and plant populations.

Increased runoff could also increase the delivery of nutrients and toxic chemicals into coastal ecosystems near urban communities. This would degrade water quality and increase the risk of harmful algal blooms. Regional fishing, hunting, and ecotourism enterprises could all be affected.

Reductions in freshwater input could also increase the salinity of estuarine systems, limiting productivity and biodiversity. Permanent reductions of freshwater flows could contribute to major reductions of biological productivity in alluvial bay systems, such as Gulf Coast lagoons.

CHANGES IN WIND PATTERNS COULD AFFECT COASTAL AND ESTUARINE CIRCULATION PATTERNS AND UPWELLING AND DOWNWELLING OF WATER IN MARINE SYSTEMS

Young organisms of many species, such as blue crab, menhaden, and bluefish, are transported into or out of estuaries by wind-driven, nearshore circulation patterns. Changed patterns would affect the normal life cycle of these species, and could diminish, if not eliminate, local populations.

In addition, wind patterns are important drivers of coastal upwelling, which provides needed nutrients to some regions. Diminution of this upwelling could reduce the ocean's productivity in these coastal areas. In contrast, increased productivity should occur in those areas that experience increased upwelling.

CHANGES IN THE FREQUENCY AND INTENSITY OF STORMS COULD INCREASE FLOODING AND THREATEN COASTAL AQUACULTURE AND FISHING INDUSTRY FACILITIES

Storm events are major drivers of coastal erosion. In addition, hurricane landfalls on the East Coast and in the Gulf of Mexico have historically been associated with significant coastal flooding. Hurricanes Dennis, Floyd, and Irene cumulatively led to

50- to 500-year floods in North Carolina during 1999. In addition to their impact on humans, these floods delivered large amounts of nutrients to the estuaries that caused oxygen depletion and harmful algal blooms.

Coastal aquaculture facilities are also highly vulnerable to the high winds and storm surges associated with coastal storms. Although the effects of climate change for storm events remain uncertain, the possibility of increased storm intensity is a significant concern.

NATURAL CLIMATE VARIABILITY, SUCH AS EL NIÑO EVENTS, RESULTS IN CHANGES IN OPEN-OCEAN PRODUCTIVITY, SHIFTS IN THE DISTRIBUTION OF ORGANISMS, AND MODIFICATIONS IN FOOD WEBS, FORESHADOWING WHAT WOULD HAPPEN IF CLIMATE CHANGE ACCELERATED

Natural climate variability exists independent of anthropogenic climate change, but may act in tandem with (or opposition to) anthropogenic climate change. The consequences are difficult to predict. Climate change could increase the frequency, duration, and/or severity of El Niño events, which have important ecological effects, heightening impacts on human society. In particular, El Niño events are often associated with mass coral bleaching, which threatens the longterm sustainability of these ecosystems.

OVER THE COMING CENTURY, CHANGES IN TEMPERATURE OR SALINITY OF NORTH ATLANTIC WATER IN THE ARCTIC MAY SLOW OR SHUT DOWN THE SLOW-MOVING THERMOHALINE CIRCULATION THAT DELIVERS COLD, DENSE, OXYGENATED WATER TO THE DEEP SEA

This would affect delivery of oxygen and nutrients from the ocean surface to the deep ocean in coming centuries, with unknown consequences for communities of deep-sea animals.

In addition, this change in circulation could alter the distribution of heat throughout the waters and atmosphere of the North Atlantic, which would affect the geographic distribution of fisheries.

It is possible that other such climate surprises could manifest in response to climate change, resulting in rapid, unpredictable changes in the marine environment.

CLIMATE-INDUCED CHANGES IN OCEAN CHEMISTRY COULD DIMINISH THE ABUNDANCE OF MICROSCOPIC OPEN-OCEAN PLANTS AND ANIMALS

Model results indicate that a doubling of the preindustrial atmospheric concentration of atmospheric carbon dioxide (currently projected to occur by the middle of the 21st century) could reduce the amount of calcium carbonate in ocean waters by 30 percent. This would limit the growth and abundance of calcium carbonate-dependent organisms. Some of these highly abundant organisms, such as diatoms and dinoflagellates, produce a chemical (dimethyl sulfide) that ultimately helps to cool surface air temperatures. Thus, changes in calcium carbonate chemistry could indirectly reinforce global warming. Our knowledge of these interactions is rudimentary, making it difficult to predict the consequences of any chemical changes.

BIBLIOGRAPHY

"As the Planet Heats up, Will Topsoil Melt Away?" *Journal of Soil and Water Conservation.* January 2004.

Conservation International. *Current and Projected Effects of Climate Change.* Available online at *http://www.conservation.org.*

———. *Weathering Climate Change.* Available online at *http://www.conservation.org.*

Intergovernmental Panel on Climate Change. *IPCC Third Assessment Report: Climate Change 2001: Synthesis Report: Summary for Policymakers.* Available online at *http://www.ipcc.ch/pub/un/syreng/spm.pdf.*

Morrison, Jim. *The Incredible Shrinking Polar Bears.* National Wildlife Federation. February/March 2003. Available online at *http://www.nwf.org.*

Moser, Susan. "Trouble in the Heartland." *Catalyst: Union of Concerned Scientists.* Spring 2003. Available online at *http://www.ucsusa.org.*

National Oceanic and Atmospheric Administration. *Science: The Antarctic Ozone Hole.* Available online at *http://www.ozonelayer.noaa.gov/.*

———. *Stratospheric Ozone: Science: Ozone Basics.* Available online at *http://www.ozonelayer.noaa.gov/.*

———. *Stratospheric Ozone: Science: Ozone Depletion.* Available online at *http://www.ozonelayer.noaa.gov/.*

Natural Resources Defense Council. *Heat Advisory: How Global Warming Causes Mores Than Bad Air Days.* July 2004. Available online at *http://www.nrdc.org/.*

Nickens, T. Edwards. *North American Fish Feel the Heat.* National Wildlife Federation. June/July 2002. Available online at *http://www.nwf.org.*

Pew Oceans Commission. *America's Living Oceans: Charting a Course for Sea Change.* May 2003. Available online at *http://www.pewoceans.org.*

Schneider, Stephen, and Terry Root. *Wildlife Responses to Climate Change.* Washington, D.C.: Island Press, 2002, pp. 19–22.

Smith, Joel B. *A Synthesis of Potential Climate Change Impacts on the United States.* Pew Center for Climate Change. April 2004, pp. 10–17. Available online at *http://www.pewclimate.org/.*

U.S. Environmental Protection Agency. *Emissions.* Available online at *http://yosemite.epa.gov/oar/globalwarming.nsf/content/emissions.html.*

———. *The U.S. Greenhouse Gas Inventory Program.* April 2002. Available online at *http://yosemite.epa/gov/oar/globalwarming.nsf/UniqueKeyLookup/ RAMR5CZKVE/$File/ghbrochure/pdf.*

World Wildlife Fund. *No Place to Hide: Effects of Climate Change on Protected Areas.* 2003. Available online at *http://www.panda.org.*

Dowie, Mark. *Losing Ground: American Environmentalism at the Close of the Twentieth Century.* Cambridge, MA: MIT Press, 1995.

Postel, Sandra. *Pillars of Sand.* New York: W. W. Norton & Company, Inc., 1999.

Quammen, David. *Song of the Dodo.* New York: Scribner, 1996.

Turco, Richard P. *Earth Under Siege: From Air Pollution to Global Change.* New York: Oxford University Press, 2002.

Wilson, E. O., ed. *Biodiversity.* Washington, D.C.: National Academies Press, 1988.

———. *Biophilia.* Cambridge, MA: Harvard University Press, 1986.

WEBSITES

Conservation International
http://www.conservation.org

Intergovernmental Panel on Climate Change
http://www.ipcc.ch/

National Oceanic and Atmospheric Administration
http://www.ozonelayer.noaa.gov/

National Wildlife Federation
http://www.nwf.org

Natural Resources Defense Council
http://www.nrdc.org/

The Nature Conservancy
http://nature.org/

Pew Center for Climate Change
http://www.pewclimate.org/

Pew Oceans Commission
http://www.pewoceans.org

U.S. Environmental Protection Agency
http://yosemite.epa.gov/oar/globalwarming.nsf/content/emissions.html

World Wildlife Fund
http://www.panda.org

INDEX

INDEX

INDEX

ABOUT THE CONTRIBUTORS

YAEL CALHOUN is a graduate of Brown University and received her M.A. in Education and her M.S. in Natural Resources Science. Years of work as an environmental planner have provided her with much experience in environmental issues at the local, state, and federal levels. Currently she is writing books, teaching college, and living with her family at the foot of the Rocky Mountains in Utah.

Since 2001, DAVID SEIDEMAN has served as editor-in-chief of *Audubon* magazine, where he has worked as an editor since 1996. He has also covered the environment on staff as a reporter and editor for *Time*, *The New Republic*, and *National Wildlife*. He is the author of a prize-winning book, *Showdown at Opal Creek*, about the spotted owl conflict in the Northwest.

Look. It's right here. See it on the bottom left of this page?

Here.

Yumi Hotta

On the bottom left of the inside flap of *JC* (Jump Comics) you'll always see this apple mark.*

Apparently it's left over from some kind of campaign that happened over a decade ago.

It's a mysterious, enduring logo. Will it be used again someday?!

—Yumi Hotta

* This refers to the apple-shaped logo that appears on the Japanese editions of *Hikaru no Go*. —Ed.

It all began when Yumi Hotta played a pick-up game of go with her father-in-law. As she was learning how to play, Ms. Hotta thought it might be fun to create a story around the traditional board game. More confident in her storytelling abilities than her drawing skills, she submitted the beginnings of **Hikaru no Go** to **Weekly Shonen Jump**'s Story King Award. The Story King Award is an award that picks the best story, manga, character design and youth (under 15) manga submissions every year in Japan. As fate would have it, Ms. Hotta's story (originally named, "*Kokonotsu no Hoshi*"), was a runner-up in the "Story" category of the Story King Award. Many years earlier, Takeshi Obata was a runner-up for the Tezuka Award, another Japanese manga contest sponsored by **Weekly Shonen Jump** and **Monthly Shonen Jump**. An editor assigned to Mr. Obata's artwork came upon Ms. Hotta's story and paired the two for a full-fledged manga about go. The rest is modern go history.

[1]

HIKARU NO GO VOL. 19
SHONEN JUMP Manga Edition

STORY BY YUMI HOTTA
ART BY TAKESHI OBATA
Supervised by YUKARI UMEZAWA (5 Dan)

Translation & English Adaptation/Naoko Amemiya
English Script Consultant/Janice Kim (3 Dan)
Touch-up Art & Lettering/Inori Fukuda Trant
Design/Julie Behn
Editor/Gary Leach

VP, Production/Alvin Lu
VP, Sales & Product Marketing/Gonzalo Ferreyra
VP, Creative/Linda Espinosa
Publisher/Hyoe Narita

Printed in Canada

Published by VIZ Media, LLC
P.O. Box 77010
San Francisco, CA 94107

10 9 8 7 6 5 4 3 2 1
First printing, May 2010

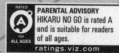

PARENTAL ADVISORY
HIKARU NO GO is rated A
and is suitable for readers
of all ages.
ratings.viz.com

www.viz.com

THE WORLD'S
MOST POPULAR MANGA

www.shonenjump.com

19 ONE STEP FORWARD!

STORY BY
YUMI HOTTA

ART BY
TAKESHI OBATA

Supervised by
YUKARI UMEZAWA
(5 Dan)

● Hikaru Shindo ●

● Shinichiro Isumi ●

● Tatsuhiko Kadowaki ●

Meet the Characters

● Kosuke Ochi ●

● Yoshitaka Waya ●

● Akira Toya ●

HIKARU NO GO

Story Thus Far

Hikaru Shindo discovers an old go board one day up in his grandfather's attic. The moment Hikaru touches the board, the spirit of Fujiwara-no-Sai, a genius go player from Japan's Heian Era, enters his consciousness. Sai's love of the game inspires Hikaru, as does a meeting with the child prodigy Akira Toya, son of go master Toya Meijin.

Upon entering junior high, Hikaru joins the school go club and his participation in a tournament makes him take the game more seriously. He rapidly improves and quits the go club, deciding to take the exam to become an insei. He passes the insei exam and as he competes against his peer rivals Waya, Isumi and Ochi, his skills develop at a surprising speed. On his first attempt to pass the pro exam, he struggles due to his inexperience but somehow manages to clear the preliminaries. To overcome his weaknesses he puts himself through intensive training at a go salon in town and gains the confidence needed to pass the pro exam. Finally on the same playing field as his rival Akira, Hikaru starts taking his first steps down the road of professional go. Amidst the crowd of formidable veterans, will he be the one to usher in the new wave of top-flight go players?

 ● Atsushi Kurata ●

 ● Toshinori Honda ●

 ● Ichiryu Kisei ●

 ● Kuwabara Hon'inbo ●

 ● Gokiso 7 dan ●

 ● Hitoshi Koike ●

 ● Akari Fujisaki ●

 ● Harumi Ichikawa ●

CONTENTS

19

"A MILLION GAMES AND I STILL DON'T GET IT."

"GO IS INFINITE."

"THE BOARD REMAINS UTTERLY DARK."

"IS THERE AN ANSWER?"

"I HAVE UNFETTERED MYSELF, BUT WILL REMAIN IN THIS LABYRINTH OF 19 ROWS FOR ETERNITY."

"I TRY TO FEEL MY WAY FORWARD IN SEARCH OF LIGHT."

"...ATTAIN THE DIVINE MOVE."

"I HAVE YET TO..."

Game 149 "The Strongest Shodan"

TRYING TOO HARD TO SAVE IT WILL ONLY DESTROY IT.

BUT SAY I WENT HERE... THEN IT'D BE BETTER NOT TO HAVE THIS EXCHANGE.

IN THAT CASE I'D GO...

RECEPTION

EH?

HI THERE, HARUMI. YOU DOIN' ALL RIGHT?

HELLO, MR. KUME. IT'S BEEN A WHILE.

ZHOOP

WHO'S THAT HE'S PLAYING?

I SEE PEOPLE ARE CROWDING AROUND THE YOUNG SENSEI.

YOU MIGHT WANT TO WATCH, MR. KUME. YOU'LL PROBABLY LEARN SOMETHING LISTENING TO TWO PROS DISCUSS THE GAME.

HIKARU SHINDO. HE'S A SHODAN IN HIS FIRST YEAR AS A PRO.

BUT IT'S ONLY BEEN RECENTLY.

YES...

¥126

¥178

SO SHINDO COMES HERE EVERY SO OFTEN?

NAH... I'M SUCH A NOVICE IT'D ALL GO RIGHT OVER MY HEAD.

12

AGAIN?

AGAIN?

UM... YOU COULD SAY THAT...

THE TWO OF THEM GET ALONG WELL?

IT'S GONNA START!

THINGS ARE HEATING UP.

TIME TO DISPERSE.

GIMME A BREAK! **YOU** DIDN'T NOTICE THIS DESCENT!

JUST, "OH, RIGHT"? YOU SHOULD HAVE NOTICED THAT YOUR-SELF, SHINDO!

AND YOU KEEP SAYING, "OH, RIGHT"!

AND **YOU** OVERLOOKED THIS ATTACHMENT!

NO, NOT THREE TIMES! FOUR!

HUH?

NOT EXACTLY.

WOW... THEY'RE ARGUING AT SUCH A SOPHISTICATED LEVEL.

I'M LEAVING!

YOU'RE FULL OF IT! I'D NEVER SAY THAT SIX TIMES!

YOU'VE BEEN COUNTING? WHAT, GOT NOTHING ELSE TO DO? WELL, YOU'VE SAID "I SEE" SIX TIMES AND ACKNOWLEDGED THE POINT I MADE!

YOUR BAG.

SEE?

JUST LIKE GRADE SCHOOL KIDS WHEN THEY FIGHT.

I AM A **MERE** 3 DAN MYSELF.

SHINDO'S A MERE SHODAN. THE YOUNG SENSEI PLAYS IN THE HON'INBO LEAGUE. FOR SHINDO TO ADDRESS HIM LIKE AN EQUAL IS INAPPROPRIATE, IF YOU ASK ME.

RANK DOESN'T NECESSARILY REFLECT ABILITY! DON'T DISMISS SHINDO JUST BECAUSE HE'S A SHODAN!

YOU HAVE TO PUT IN YOUR TIME AND MOVE UP THE RANKS RUNG BY RUNG.

NO MATTER WHAT YOUR ABILITY, EVERYONE STARTS AS A SHODAN.

YUP.

THE TWO OF THEM **WERE** FIGHTING, RIGHT?

WHO'RE YOU PLAYING TODAY?

JAPAN GO ASSOCIATION

40 METERS AHEAD

...THIRD ROUND OF THE FIRST PRELIMS FOR THE HON'INBO TOURNAMENT.

KAWASAKI 3 DAN, IN THE...

MAN, I WISH I COULD PLAY AGAINST THE HIGHER RANKS. THE FIRST YEAR AS A PRO IS NO FUN, HUH?

YEAH? I'M PLAYING IN THE SECOND ROUND OF THE FIRST PRELIMS FOR THE MEIJIN TOURNAMENT, AGAINST TOYAMA 2 DAN.

LISTEN, EVEN TOYA HAD TO START AT THE SAME EXACT POINT AS YOU AND ME!

I'M NOT. IT'S JUST THAT TOYA GETS TO GO UP AGAINST 9 DAN PLAYERS.

HEY! YOU CAN'T GO INTO THIS GAME TAKING KAWASAKI 3 DAN SO LIGHTLY!

ZHOOP

NOTHIN' TO BE GAINED BY BEING IN A HURRY.

YOU'RE A SHODAN. I'M A 2 DAN.

EVEN FOR THE OTEAI MATCHES, IF YOUR RANK DOESN'T GO UP YOU'LL NEVER FACE A REALLY STRONG PLAYER.

FOR ANY TOURNAMENT, WE HAVE TO WIN OUR WAY THROUGH THE PRELIMS WHERE THE LOWER RANKED PLAYERS FROM SHODAN TO 4 DAN PLAY EACH OTHER.

IT'S JUST SO DARNED SLOW...

I KNOW, BUT...

HEY, WAYA, YOU ORDERING IN A LUNCH TODAY?

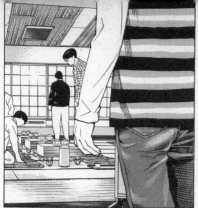

NAH... I LIKE TO GET OUT AT LUNCH.

I FEEL LIKE HAVING THE PORK CUTLET OVER RICE HERE.

YOU?

SO, WHERE DO I PLAY?

TODAY I'M PLAYING SHINDO SHODAN. I THINK HE JUST WENT PRO THIS YEAR. IT'S BEEN SIX MONTHS SINCE THEN...

...AND ANYONE REALLY GOOD WOULD BE 2 DAN BY NOW.

I WON'T HAVE TO BE TOO CAUTIOUS, THEN.

GOOD MORNING.

OH...

MORN-ING.

HA HA...

LAST TIME I HAD OCHI NEXT TO ME.

HEY, SHINDO, WE'RE NEIGHBORS TODAY.

BUT I SHOULDN'T LET DOWN MY GUARD.

YOU BET.

LET'S DO OUR BEST.

BEEEEP

I WANT TO PLAY BLACK.

...

KSHH

TWO... FOUR...SIX... EIGHT...

...FIF-TEEN.

KTNK

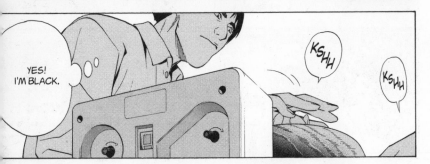

YES! I'M BLACK.

KSHH

KSHH

ONEGAI-SHIMASU.

TOP O' THE WORLD... HEH...

KLNK

I WOKE UP IN A GOOD MOOD TODAY.

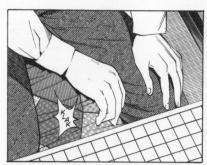

22

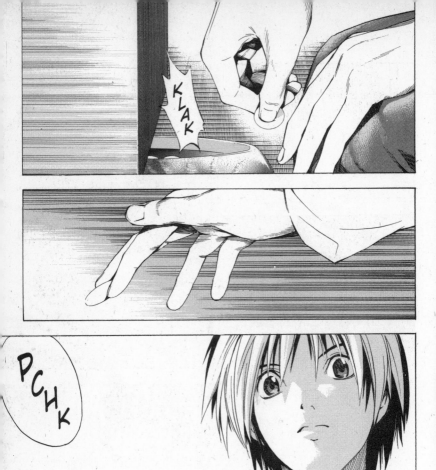

BEEEEEP

SEPLTURA

KCHK

KLAK

KCHK

KCHK

KLAK

TIME
FOR
LUNCH.

24

BUSTLE
BUSTLE

...

NOW FOR MY PORK CUTLET OVER RICE.

WHY IS HE A SHODAN?

HEY!

HUH?

...ALL SPRING AND SUMMER?

UM... DID YOU HEAR ABOUT THAT GUY WHO DIDN'T SHOW UP FOR HIS GAMES...

WHY IS SHINDO STILL A SHODAN?

OH...

!

THAT GUY'S GONNA BE A SHODAN FOR A WHILE.

...BEFORE HE MOVES UP.

THAT WAS SHINDO. HE RACKED UP A LOT OF FORFEITS, SO HE'S GOT A LOTTA GAMES TO MAKE UP...

YEAH, I HEARD A RUMOR...

BUT HE'S SO STRONG...

SHODAN FOR A WHILE, HUH?

HE'S THE STRONGEST SHODAN EVER.

WE'RE THROUGH WITH DIAPERS. LIFE AS A MOM IS FINALLY GETTING A LITTLE EASIER.

I'M SO OUT OF SHAPE. I ONLY PLAY ONCE A MONTH.

INCREDIBLE TIMING...

SHINDO...

...WAS TO CONNECT, BUT HE WASN'T CONTENT WITH THAT, SO HE ATTACHED. WHAT A MOVE...

THE OBVIOUS RESPONSE TO BLACK'S CUT...

HEY, I HEAR THERE'S GONNA BE A NEW TOURNA-MENT FOR YOUNG PROS.

GUESS IT'S ABOUT TIME.

...

OH YEAH?

YOUNG PROS?!

YEAH, FOR THOSE 18 AND UNDER. IT'S CALLED SOMETHING LIKE THE JAPAN, CHINA AND KOREA TEAM TOURNAMENT.

A JAPAN, CHINA AND KOREA TEAM TOURNA-MENT?!

EIGHTEEN AND UNDER?!

SEPULTURA

EIGHTEEN
AND UNDER?!
FOR REAL?!

CHINA?!
KOREA?!
INCREDIBLE!

LOW-RANKED PLAYERS LIKE US WILL GET OUR CHANCE TO PLAY ON THE BIG STAGE!

SHINDO'S GONNA LOVE THIS!

BUT HOW WILL THEY DECIDE WHO'LL BE ON JAPAN'S TEAM?

I BET THEY'LL HAVE A QUALIFYING TOURNAMENT OR SOMETHING.

! WAYA...

SAY, WHAT'S UP?

...

WELL, **WE** DON'T HAVE TO WORRY ABOUT IT. HA HA...

HEY...

WHAT'S THAT?

THIS?

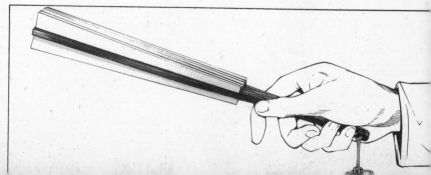

I JUST BOUGHT IT AT THE SHOP.

SEPULTURA

...AND GET TODAY'S MATCH WRAPPED UP.

WELL, I'D BETTER GO...

WHAT IS IT?

...

SEPULTURA

BEEEP

Game 150 "A Big Stage"

DECEMBER

...THE DIFFERENCE IN THEIR RANKS.

ICHIRYU SENSEI APPEARS RELAXED. I SUPPOSE IT'S...

TOYA SEEMS DETERMINED TO KEEP UP WITH ICHIRYU SENSEI.

MAY I TAKE A LOOK?

BUT NOT THIS AKIRA TOYA.

I KNOW ICHIRYU SENSEI WELL.

NOGI SENSEI.

SO SERIZAWA CAME TO WATCH TOO.

AS ONE OF MY HON'INBO LEAGUE RIVALS, I'D BETTER LEARN MORE ABOUT HIM.

MMM...

KLAK

...

WHAT WILL TOYA DO?

IF BLACK JUMPS AND WHITE UNEXPECTEDLY ATTACHES, THEN WHITE WILL BE ATTACKED.

ICHIRYU IS MAKING SOLID, SECURE MOVES.

IF THINGS GO THE SAME WAY TODAY, THEN TOYA'S NO ONE I HAVE TO WORRY ABOUT.

TOYA SEEMED UNABLE TO MUSTER THE STRENGTH HE NEEDED IN HIS GAME WITH ZAMA SENSEI.

I SUPPOSE I SHOULD BE GRATEFUL.

IT'S BEEN AGES SINCE I'VE SEEN A NEW PRO POSE A REAL THREAT.

MAYBE A FRIEND OF TOYA'S CAME TO WATCH.

AN INSEI...OR A PRO?

LOOK AT THIS GAME... EVEN IF WHITE CAN CAPTURE A STONE AFTER ATTACHING AND PULLING BACK, HE'LL STILL BE ATTACKED. THAT'S A TOUGH SPOT.

...

LET'S SEE HOW HE'S PLAYING...

...

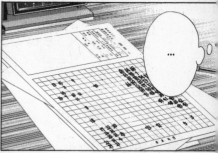

...

TOYA!

WAIT... WHAT ABOUT ATTACHING TO THE CORNER ENCLOSURE?

KCHK

KURATA 7 DAN....PLAYED IN KOREA'S SANSEI INSURANCE CUP...

...LOSING A...CLOSE GAME TO AHN DAESUN OF KOREA.

TAP TAP

YES, SIR.

SECONDS!

I'M FLYING BACK TO JAPAN WITH A HEAVY HEART.

NO FUN FOR THE PUBLISHING DEPARTMENT TO WRITE UP A LOSS, EH?

NOT WHAT *GO WEEKLY* READERS WANT TO KNOW.

KURATA RELIEVES HIS FRUSTRATION BY PIGGING OUT ON HIS FLIGHT BACK TO JAPAN.

THAT'S HIS SECOND TIME ASKING FOR SECONDS.

AS HIS SUCCESSOR, I HAD NO IDEA WHAT WAS WHAT. TRUTH IS, BEING IN THE KNOW IS NOT MY THING, WHAT WITH HAVING TO NETWORK AND MAKE CONNECTIONS AND ALL. HA HA...

IT SURE WAS A SUDDEN TRANSFER.

HAVE YOU GOTTEN USED TO THE PUBLISHING DEPARTMENT, KOSEMURA? MUST BE HARD TO TRY TO FILL AMANO'S SHOES.

ONLY HALF A YEAR AGO IT WAS LIKE, JAPAN HAS KOYO TOYA... NUFF SAID, END OF DISCUSSION.

IT LOOKS LIKE THE INTERNATIONAL TOURNAMENTS ARE REALLY STARTING TO CENTER AROUND KOREA. A SIGN OF THE TIMES, I SUPPOSE.

GLARE

KURATA'S NOT AT THAT LEVEL.

SHH...

KURATA IS...

...

HE SAYS I NEED TO DO BETTER, HUH?

AGH! THAT JERK AHN DAESUN!

HE SAYS KOREANS CALL ME THE AHN DAESUN OF JAPAN, SO MY LOSSES...

SCARF

...SULLY HIS NAME, HUH? THE NERVE!

SCARF

SO, KURATA, ARE YOU NOT FEELING WELL?

HE SHOULD BE CALLED THE KURATA OF KOREA!

IT IS TOO!

THAT MEANS YOU'RE WELL KNOWN IN KOREA, KURATA. THAT'S NOT SUCH A BAD THING.

CHOMP CHOMP

SCARF SCARF

MUNCH MUNCH

YOU'VE GOT SPIRIT...

BEATS ME WHY THEY'RE DOING IT.

OH, YOU MEAN THE ONE HOKUTO COMMUNICATION SYSTEMS IS SPONSORING?

...JAPAN-CHINA-KOREA JUNIOR TEAM TOURNAMENT, I THINK IT'S CALLED?

HEY, MR. DOI, DO YOU GUYS IN PR KNOW ABOUT THIS...

BUT FOR JAPAN, AKIRA TOYA'S THE ONLY ONE WHO'S MAKING ANY KIND OF IMPRESSION. HE MUST BE PLAYING ICHIRYU SENSEI RIGHT ABOUT NOW IN THE HON'INBO LEAGUE MATCH.

...HONG SUYONG...

KOREA IS TEEMING WITH SKILLED PLAYERS IN THEIR TEENS. THE TOP ONE IS KO YONG HA BUT THERE'S ALSO KIM SANG RYUL, CHANG SUNGHO, HONG SUYONG...

YET HIS NAME KEEPS GETTING BANDIED ABOUT.

THE KID WHO FORFEITED ALL THOSE GAMES, REMEMBER? NOT MUCH OF A RECORD AS AN AMATEUR, EITHER.

SHINDO SHODAN?

BESIDES HIM... I'VE HEARD SHINDO SHODAN MENTIONED.

OH!

HONG SUYONG ASKED ME SOMETHING WHILE WE WERE OVER THERE. IN JAPANESE!

WHAT WERE YOU JUST REMINDED OF?

ASKED YOU...?

POP

YOUR CONVER- SATION JUST RE- MINDED ME.

I THOUGHT YOU WERE SLEEPING OFF ALL THAT INTER- PRETING.

OF WHAT?

VOOOM

IF HIKARU SHINDO HAD REALLY GONE PRO.

48

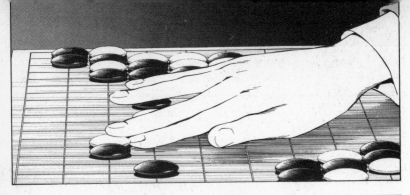

SNAP

!

TOYA...

IF HE PLAYS ELSEWHERE, BLACK'S TERRITORY WILL BE WIPED OUT, SO HE MUST RESPOND.

THAT...IS A GOOD PLAY.

THAT DIDN'T OCCUR TO ME.

NOR TO ICHIRYU SENSEI. HE'S NOT HAPPY.

HE SAW EXACTLY WHAT I SAW.

HE WON THAT ONE, BUT THINGS COULD STILL GO IN ANY DIRECTION. EVEN A SMALL MISSTEP WOULD BE FATAL.

!

AAGH!

KLAK

KSHH

I WON'T JUDGE AKIRA TOYA YET! NO, NOT UNTIL THIS MATCH IS OVER!

YOU MUST WIN THROUGH TO CLAIM TRUE STRENGTH.

...AND TECHNIQUE. THEY BOTH HAVE TO RESIST PANIC, WHICH WILL FAVOR TOYA.

FROM HERE ON OUT IT'S NOT JUST A BATTLE OF STRENGTH...

FWSH

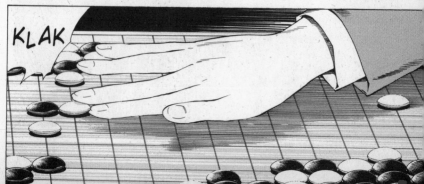

KLAK

KLAK

FWSH

WHOA! INSTANT PLAY!

FWSH

IT'S OVER.

!

▽ 12345678 △

THUNK!
RATTLE!
SLAM!

ICHIRYU SENSEI!

HEY!

WHACK

ZHOOP

OH...

THAT MOMENT WAS THE TURNING POINT.

SERIZAWA SENSEI...

IT'S QUITE FRUSTRATING TO LOSE BY A MISREAD.

NEITHER I NOR NOGI SENSEI SAW...

BUT ICHIRYU SENSEI WASN'T THE ONLY ONE WHO MISSED IT.

...THAT MOVE COMING. NOBODY IN THAT ROOM DID, EXCEPT TOYA.

HE FOUGHT ALL THE WAY TO GAME'S END.

...AND TOYA'S INNER STRENGTH THAT RESISTED THAT PRESSURE.

IT WAS ICHIRYU SENSEI'S PRIDE THAT MADE HIM TRY TO PRESSURE HIS OPPONENT...

IT WASN'T JUST THAT MOVE...

...PLAY LIKE THAT TOO!

I WANT TO...

JAPAN GO ASSOCIATION

...DOING IN THE LOWER RANKS? I CAN'T BEAT HIM.

WHAT'S A KID AS STRONG AS THIS...

...

TAP

I RESIGN.

...

GASP

THANK YOU FOR THE GAME.

THANK YOU FOR THE GAME.

KTNK

SIGH... WHAT A FAR CRY FROM THE GAME BETWEEN TOYA AND ICHIRYU SENSEI.

WELL, I'VE CLEARED THE FIRST MATCH OF THE FIRST ROUND OF PRELIMS FOR THE KISEI TOURNAMENT...

KTNK

SHINDO!

AH!

YEAH?

HEY, THERE'S SOMETHING I WANT TO ASK YOU.

YOUR GAME'S OVER ALREADY? THAT WAS QUICK.

WHEN I WAS AN INSEI I PLAYED A GAME AGAINST HIM HERE IN JAPAN. I WON!

SO YOU DO KNOW HIM.

DO YOU KNOW A KOREAN PRO NAMED HONG SUYONG?

I'M KOSE-MURA, FROM PUB-LISHING.

HE WENT PRO?!

SUYONG!

BOY, WAS HE MAD!

HEH HEH...

YOU DID?! YOU BEAT HIM?!

MR. DOI WAS PRETTY PESSIMISTIC, SAYING JAPAN DIDN'T HAVE A CHANCE OF WINNING.

THAT MEANS JAPAN DOES HAVE SOME PROMISING PLAYERS AFTER ALL!

GREAT!

THE JAPAN-CHINA-KOREA JUNIOR TEAM TOURNA-MENT?!

?

WHEN THE JAPAN-CHINA-KOREA JUNIOR TEAM TOURNAMENT FOR THE HOKUTO CUP HAPPENS THIS MAY...

...JAPAN HAS A REAL SHOT AT WINNING!

...AND YOU, WHO BEAT HONG SUYONG.

BUT WE'VE GOT TOYA, WHO BEAT ICHIRYU SENSEI...

A WORD ABOUT HIKARU NO GO

THE STONES IN THE IGO STARTER BOX

THE IGO STARTER BOX (2,980 YEN PLUS TAX) IS A
SET CONSISTING OF A GO TEXTBOOK, A NINE-BY-NINE
BOARD AND A SET OF STONES. THE CHARACTER
STONES ABOVE ARE PART OF THAT SET.

DAI NIPPON PRINTING COMPANY PROUDLY BOASTS
THAT THEY CAN PRINT ON ANYTHING EXCEPT AIR, BUT
APPARENTLY EVEN THEY HAD A HARD TIME PRINTING
ON THESE GO STONES.

SUCH SUPER-CUTE GO STONES TOO! ♡
THOUGH WHEN YOU USE THEM IT'S KIND OF
HARD TO SEE HOW THE GAME IS GOING.

THE JAPAN-CHINA-KOREA JUNIOR TEAM TOURNAMENT?!

A TOURNAMENT BETWEEN JAPAN, CHINA AND KOREA?!

SOMEONE FROM THE PUBLISHING DEPARTMENT TOLD ME ABOUT IT.

...UNDER 18, WITH THREE ON A TEAM.

YEAH! SOUNDS PRETTY INTERESTING, HUH? IT'S FOR PROS...

IT'S FOR PROS 18 AND UNDER, SHINDO SAID...

MAYBE I COULD TRY FOR IT! I JUST PASSED THE PRO TEST THIS YEAR!

I HEARD ABOUT IT TOO. IN APRIL THERE'LL BE QUALIFYING MATCHES TO DETERMINE THE TEAM PLAYERS.

RIGHT... RIGHT...

SIGH

KSHH

...

HEY, YOU GUYS REMEMBER SUYONG?

OF THIS YEAR'S NEW PROS, I'M AFRAID YOU AND KADOWAKI ARE BOTH OUT OF IT. BUT HONDA'S JUST 18, RIGHT?

KSHH

HE'S ALREADY PLAYING THAT HIGH?

A 9 DAN?

HE'S NOW A PRO. AND HE BEAT A 9 DAN NAMED SOMETHING OR OTHER.

THE KOREAN KID WE MET AT THE GO SALON?

OF COURSE, THEIR TOURNAMENT SYSTEM IS DIFFERENT FROM OURS.

I HEAR THAT IN KOREA TOP PLAYERS CAN BE MATCHED WITH SHODAN FROM THE VERY FIRST ROUND.

AKIRA TOYA AND ME!

THE KOREAN TEENS ARE AMAZING. KO YONG HA'S JUST 16 AND ALREADY A TOP PLAYER.

IT'S THE SAME IN CHINA.

THOUGH I SUPPOSE JAPAN HAS AKIRA TOYA.

KSHH

JUST LIKE SUYONG AND TOYA!

IF I WIN, I'LL MOVE TO THE SECOND PRELIMS AND FINALLY GET TO COMPETE WITH HIGH-RANKED PLAYERS!

MY NEXT MATCH IS IN THE FINALS OF THE HON'INBO FIRST PRELIMINARIES!

I'M NOT GONNA TRAIL TOYA FOREVER, YOU CAN BET ON THAT!

YOU'RE NOT THE ONLY ONE.

...

I'M HEADING FOR THE TOP TOO!

WAYA...

BUT MORISHITA SENSEI SAYS I'LL GET BETTER. AND I WILL!

SO FAR I FEEL LIKE I'VE BEEN TREADING WATER, ESPECIALLY COMPARED TO YOU, SHINDO.

MY LIMIT'S NOT EVEN IN SIGHT YET!

GEEZ, WAYA...

67

YOU'VE GOT THE FIRE, WAYA.

THERE YA GO...

I *WILL* MAKE THE TEAM FOR THIS NEW TOURNAMENT!

REMEMBER THAT CHINESE KID I TOLD YOU ABOUT, LE PING...

THAT'S RIGHT!

YO! WE'RE HERE!

I'M SERI-OUS!

QUIT KIDDIN' AROUND!

...WHO LOOKS JUST LIKE YOU? IT'D BE A HOOT IF THE TWO OF YOU QUALIFIED AND ENDED UP PLAYING EACH OTHER!

JAPAN GO ASSOCIATION

I'M A LITTLE BIT AHEAD AT THIS POINT...

KLAK

KLAK

...

WELL, THAT'S AS FAR AS HE'LL GO!

HE'S ONLY A SHODAN, BUT HE MADE IT TO THE FINALS OF THE FIRST PRELIM.

KCHK

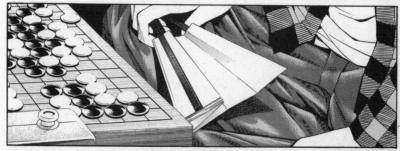

70

OKAY...
I'M IN GOOD
SHAPE...

WHAT?
OH NO...

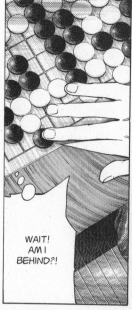

WAIT!
AM I
BEHIND?!

WHAT
INCREDIBLE
ACCURACY!

THERE'S
PRACTI-
CALLY NO
ERROR IN
HIS END-
GAME.

HE... OVERTOOK ME.

HE BEAT ME BY...ONE AND A HALF POINTS.

THANK YOU FOR THE GAME.

WHAT A COMPLETE AND UTTER MESS! YOU LOST, OBVIOUSLY.

THIS WAS THE GAME YOU PLAYED THIS WEEK?

I WON BY ONE AND A HALF POINTS!

HAH! I CAME FROM BEHIND AND BEAT HIM!

HMPH!

IF HE HADN'T, HE'D HAVE SHUT YOU DOWN WELL BEFORE THE ENDGAME. YOUR OWN PLAYING WAS ATROCIOUS.

YOU ONLY WON BECAUSE YOUR OPPONENT MISREAD HIS POSITION.

SKOOT

SKOOT

LIKE YOU WOULD'VE DONE BETTER!

WHAT?!

IN A BLINK! AND YOU KNOW IT!

SEE HERE... HERE... AND HERE? NOTHING TO IT.

KCHK
KCHK
KLAK

KLNK
KLNK

I'D HAVE STOPPED YOUR COMEBACK AT THIS POINT.

...

A COMEBACK WOULD BE IMPOSSIBLE.

WITHOUT TERRITORY IN THE CENTER, BLACK'S GOOSE IS COOKED.

PCHK PCHK
PCHK

BUT THEN... WON'T YOU LEAVE A GAP HERE?

...

TRY TO DANCE AROUND THAT!

WAIT! I WOULD CHANGE IT LIKE THIS!

KLAK KLAK
KLAK
KLAK

NOBODY COULD FIGHT IT OUT FROM THIS POSITION!

I COULD PLAY A DIAGONAL HERE AND THREATEN WHITE'S TERRITORY...

IF YOU DON'T GET ANYTHING IN THE CENTER, YOU'D FALL EVEN FURTHER BEHIND IN TERRITORY AND... NO COMEBACK.

OH YEAH? I COULD! THE DIAGONAL'S ENOUGH!

YOU WON'T BE ABLE TO ACT SO SUPERIOR FOR MUCH LONGER. SOMEDAY I'LL BEAT YOU IN AN OFFICIAL MATCH.

KSHH

KSHH

YEP!

PSST PSST

EVERY DAY THEIR FIGHTING GETS FIERCER...

GOT THAT RIGHT!

YOU REFUSE TO GIVE IN.

HEY, WAIT...

WHEN **WILL** I EVER GET TO PLAY YOU?

KSHH

KSHH

'COURSE, THERE ARE TITLE TOURNAMENTS THAT HAVEN'T EVEN STARTED PRELIMS.

MAYBE I'LL GET TO PLAY YOU THERE!

...WILL DECIDE WHO REPRESENTS JAPAN!

THE HOKUTO CUP! YOU'VE HEARD ABOUT THAT NEW TEAM TOURNAMENT, RIGHT? THE QUALIFYING MATCHES...

KSHH

THE HOKUTO CUP...

I WON'T BE IN THE PRELIMS.

KSHH

I ALREADY HAVE A SPOT ON THE TEAM.

KSHH

HARDLY! IT'S CALLED BEING SEEDED, SHINDO.

NO FAIR!

WHAT?! **YOU** GET TO SKIP THE QUALIFIERS?!

AKIRA'S IN THE HON'INBO LEAGUE, AFTER ALL.

INDEED.

YEP, THAT'S HOW IT GOES.

REMEMBER WHERE ALL THOSE FORFEITS GOT YOU? DEAL WITH IT!

THAT'S RIGHT. HE'S GOT BETTER THINGS TO DO THAN PLAY QUALIFIERS.

OKAY, I GET THE POINT. RANK HAS ITS PRIVILEGES, BUT, Y'KNOW, PRIVILEGES...

...

EH?

KSHK

...AREN'T THE SAME AS ABILITY.

SKOOT

HEY! QUIT DELUDING YOURSELF!

I'LL GET THROUGH THE QUALIFYING TOURNAMENT IN APRIL, YOU'LL SEE.

UNTIL THEN, I'M OUTTA HERE.

HOW DARE YOU ACT LIKE YOU'RE EQUAL TO THE YOUNG MASTER!

MR. KITA-JIMA!

TNK

MARK IT! THAT'S WHEN I'LL JOIN YOU AS A MEMBER OF JAPAN'S TEAM.

THAT'S FOUR MONTHS FROM NOW, TOYA.

SHINDO...

...AND
KEEP
ADVANC-
ING...

ZHOOP

I'LL TAKE IT
ONE STEP
AT A TIME...

...UNTIL
I ATTAIN
THE DIVINE
MOVE.

Game 152: "The Opponent Is a 7 Dan"

HEY, HARUMI! HAPPY NEW YEAR!

Game 152 "The Opponent Is a 7 Dan"

HOW'S THE YOUNG MASTER? HAVEN'T SEEN HIM AROUND MUCH SINCE SHINDO STOPPED COMING.

HAPPY NEW YEAR TO YOU!

HEY, MR. KITAJIMA! GOOD TIMING. HOW ABOUT A GAME?

CHINESE CLASS?

HE SAID HE'D COME IN TODAY ON HIS WAY HOME FROM HIS CHINESE CLASS.

ZHOOP

IT'S NOT JUST CHINESE. HE'S STUDYING KOREAN TOO!

EH?

AH... SPEAK OF THE DEVIL.

PRETTY IMPRESSIVE. BUT I SHOULDN'T BE SURPRISED.

YEAH, AND IT'S TOUGH. THEY GET ALL MIXED UP IN MY HEAD.

THANKS.

I'LL MAKE YOU SOME COFFEE.

SO, YOUNG MASTER, YOU'RE TAKING CHINESE AND KOREAN LESSONS?

SHINDO WAS SURE CONFIDENT HE'D MAKE THE TEAM.

SPEAKING OF THAT TOURNA- MENT...

IT'S FOR THE JAPAN-CHINA- KOREA JUNIOR TEAM TOURNAMENT, RIGHT?

...FOR THEM TO BE MUCH USE TO ME FOR THAT. IT'S IN MAY, AFTER ALL.

I WON'T LEARN EITHER WELL ENOUGH...

OR SOMETHING LIKE THAT. WHO DOES HE THINK HE IS?!

"I'M GONNA ATTAIN THE DIVINE MOVE"?!

HE'S ALL TALK! WHAT'S THAT HE SAID AS HE WALKED OUT?

IT WAS SHINDO'S RESOLVE AS A PROFESSIONAL PLAYER THAT MADE HIM SAY THAT.

RIGHT, AKIRA SENSEI?

EVERYONE HERE IS A FAN OF YOURS!

PASS OVER YOU?! I DON'T WANNA HEAR YOU SAY THAT!

MR. KITA-JIMA...

IF I DON'T PAY ATTENTION, HE MIGHT PASS RIGHT OVER ME.

I'D BETTER WORK HARD TOO.

THAT'S RIGHT.

DON'T WORRY...

89

...AND KEEP ADVANCING.

I, LIKE SHINDO, WILL TAKE IT STEP BY STEP...

HEH...

THAT'S WHAT I LIKE TO HEAR.

GOOD.

NOW SIT DOWN, MR. KITAJIMA, SIT DOWN...

I DON'T SEE IT.

IS SHINDO REALLY THAT GOOD?

THINK SO? WHATEVER HE'S GOT, IT CAN'T COMPARE TO THE YOUNG MASTER. **HE'S** EVEN BEATEN ICHIRYU KISEI!

SHINDO HAS A MATCH AGAINST A 7 DAN THIS WEEK. THAT'LL...

...SHOW US WHAT HE'S GOT.

HMPH!

YOU'RE BEING AWFULLY PARTIAL TO AKIRA, MR. KITAJIMA.

THE JAPAN-CHINA-KOREA JUNIOR TEAM TOURNAMENT

N-CHINA-KOREA
AM TOURNAMENT!

THE HOKUTO CUP

China-Korea
Tournament The Hokuto Cup

THE HOKUTO CUP

The Japan-China-Korea
Junior Team Tournament   The Hokuto Cup

5/4–5/5   Location: Seven Star Hotel
          Sponsored by: Hokuto Communication Systems
          With the support of: Japan Go Association
                    Kansai Go Association
                    Korean Go Association
                    Chinese Go Association

MR. KITAJIMA!

WHO KNOWS IF SHINDO WILL EVEN PASS THE QUALIFIERS FOR THE HOKUTO CUP?

JAPAN GO ASSOCIATION

GOOD MORNING, SHINDO.

YOU'VE AN IMPRESSIVE RECORD OF LATE, SHINDO.

FIRST TIME TO PLAY ON A THURSDAY—THAT IS, TO PLAY A HIGH-RANKING PRO, EH?

GOOD MORNING.

GOOD
MORNING.

SHINODA
SENSEI.

GOOD
MORNING.

GOOD
MORNING.

MORNING.

MORISHITA
SENSEI'S
HERE TOO.

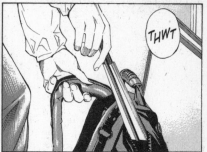

THWT

AKIRA'S ALREADY A REGULAR ON THURSDAYS.

...I'VE BEEN LOOKING FORWARD TO FOR SO LONG...

TOO BAD MY FIRST GAME, THE ONE...

FINALLY, I GET TO PLAY PROS RANKED 5 DAN AND UP.

AND I'M UP PAST THE FIRST ROUND OF PRELIMS WHERE 4 DANS AND LOWER PLAY.

...HAS TO BE AGAINST THE LIKES OF YOU.

!

THAT RANK SURPRISES ME.

GOKISO 7 DAN, HUH?

THAT KID FROM A YEAR AGO...

YOU...

YOU'RE SHINDO? MY OPPONENT TODAY?

SO YOU WERE A **PRO**...

DON'T THINK YOU KNOW HOW WELL I REALLY PLAY, SHODAN.

...BUT THAT'S BECAUSE I WENT EASY ON YOU, THINKING YOU WERE A REGULAR KID.

WELL, I MAY HAVE LOST LAST TIME...

...

RANKS ARE FOR LIFE, NO MATTER HOW WEAK YOU GET OR HOW OFTEN YOU LOSE ONCE YOU REACH THEM.

YOUR RANK DOESN'T MEAN YOU'RE ANY GOOD!

BUT IT DOES MEAN YOU OWE ME RESPECT, KID.

RESPECT?! FOR YOU?! YOU'VE GOTTA BE KIDDING!

YOU WERE **SHARP** TO SPOT THE SHUSAKU FORGERY.

HAW!

HMM...

DON'T TELL ME YOU'RE STILL AT IT.

I REFUSE TO LOSE TO YOU.

I WON'T LOSE TO YOU!

BEEEP

THE PR DEPARTMENT DID A NICE JOB ON THESE POSTERS.

THE JAPAN-CHINA-KOREA
JUNIOR TEAM TOURNAMENT

THE HOKUTO CUP

China-Korea
Tournament

The Hokuto Cup

Location: Seven Star Hotel
Sponsored by: Hokuto Communication Systems
With the support of: Japan Go Association
Kansai Go Association
Korean Go Association
Chinese Go Association

BOSS...

...WHEN IT'S ONE-ON-ONE.

TRUE, IT DOESN'T EXACTLY BOIL OVER WITH ACTION, ESPECIALLY...

GO, THOUGH... IT'S NOT REALLY VERY EXCITING.

ALL TO CELEBRATE OUR COMPANY LISTING ON THE STOCK EXCHANGE, RIGHT?

...AN **INTERNATIONAL** TEAM TOURNAMENT, TO BE PRECISE. WHEN IT'S JAPAN VS. CHINA, OR JAPAN VS. KOREA, EVEN THE AVERAGE PERSON'S INTEREST WILL BE PIQUED. THAT'S THE IDEA, ANYWAY.

THAT'S WHY WE MADE IT A TEAM TOURNAMENT...

TAP

TAP TAP

WHEN YOU PIT ONE COUNTRY AGAINST ANOTHER, LIKE WITH THE OLYMPICS, EVERYONE GETS PASSIONATE!

GO JAPAN! GO JAPAN!

AND IT'S A GOOD ONE!

GO IS ONLY MILDLY POPULAR IN JAPAN, BUT...

OUR COMPANY WANTS TO EXPAND INTO ASIA, SO THE GOAL WITH THIS PROMOTION IS TO STRENGTHEN OUR CONNECTIONS THERE.

IN FACT, IT'LL BE BETTER IF WE LOSE.

JAPAN DOESN'T EVEN HAVE TO WIN.

HUH?

...AND NOT ME, PART OF THE EXECUTIVE COMMITTEE.

NOT OUR PRESIDENT...

FACT IS, WE DON'T GIVE A HANG ABOUT GO.

THAT, IN A NUTSHELL, IS WHY WE'RE SPONSORING THIS.

INTERNATIONAL GO TOURNAMENTS GET WIDE COVERAGE IN THE MAINSTREAM MEDIA.

...IN CHINA AND KOREA IT'S LIKE A MAJOR COMPETITIVE SPORT.

WELL... HMPH!

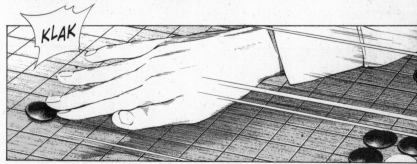

KLAK

KCHK

FWSH

KCHK

KLAK

I MAY NOT BE
THE PLAYER
I WAS, BUT MY
TECHNIQUE'S
STILL
SOUND.

...

103

I HAVE TO EXPAND BLACK'S INFLUENCE... AND WHEN HE INVADES...

I NEED TO BE CAREFUL ABOUT THE POSITION IN THE CENTER AS I PLAY.

...YOU PLAYED BACK THEN.

...AND I'M NOT THE SAME OPPONENT...

IT'S BEEN A YEAR...

NEED I REMIND YOU NOT TO UNDER-ESTIMATE ME?

I'M NOT THE SAME OPPONENT YOU PLAYED BACK THEN EITHER.

WELL, TOO BAD FOR YOU!

THAT INSULT BURNS ME UP NOW AS MUCH AS IT DID THEN! MORE, EVEN!

YOU INSULTED SHUSAKU WITH THAT FORGED SIGNATURE!

YOU'RE ON.

FINE! WE'LL SETTLE THIS ON THE BOARD.

HIKARU NO GO

STORYBOARDS

㊹

YUMI HOTTA

HIKARU NO GO 2 FOR GAME BOY ADVANCE!

I'M OBSESSED!

Obata is probably working around the clock on the coloring right now.

I'M SORRY, I'M PLAYING A GAME.

THE STORY IS THAT YOU'RE AN INSEI TAKING THE PRO EXAM TOGETHER WITH HIKARU AND HIS FRIENDS!

Just like with Hikaru No Go 1, the characters say lines that fit them perfectly. It's just so much fun.

The drawings are good and the mini-characters that move around are cute! Ooh...it's so well done.

THE INSEI INSTRUCTOR GIVES PROBLEMS THAT ARE SURPRISINGLY HARD.

I THINK I'D RECOMMEND 1 TO BEGINNING GO PLAYERS.

But with 2 you get voices in the free play mode!

It's amazing! The voice actors are so good!

THE CPU STILL TAKES A LONG TIME TO THINK, THOUGH.

Quit playing while we eat!

I'm not playing.

I'm just waiting.

Husband

Game
153
"One
Step
Forward!"

KCHK

KLAK

KLAK

I DIDN'T EXPECT HIM TO GO FOR SUCH THICKNESS.

I THOUGHT HE'D INVADE RIGHT AWAY.

WHAT'S THIS?

IS HE TRYING TO PROVOKE ME WITH A WIDE STANCE?

KLAK

KLAK

KLAK

IF HE EXTENDS HIS INFLUENCE ANY FARTHER IT'LL BE A NUISANCE FOR ME.

KLAK

KLAK

ALL RIGHT, SHALL WE START FIGHTING NOW?

KLAK

STILL BIDING HIS TIME, HUH?

KLAK

HE WAS ONLY FEISTY UNTIL THE GAME STARTED.

HMPH!

KLAK

A CROSS
ATTACH-
MENT?!

WHAT?!

KLAK

KCHK

THEN I'LL SACRIFICE
THOSE FOUR WHITE
STONES AND PUSH
INTO THE CENTER.

I HAVE QUITE A BIT
OF TERRITORY,
BUT THIS KID
SEEMS TO THINK
HE CAN ATTACK
THAT WAY.

HE THINKS
SURROUNDING
IS ENOUGH?

HMPH!

KLAK

I'LL TEACH HIM THAT HIS POSITIONAL JUDGMENT IS UTTERLY WRONG.

KLAK

CHK

KLAK

KLAK

KLAK

AN ELEGANT MANEUVER, DON'T YOU THINK?

IF WE HAD AN AUDIENCE, THEY WOULD BE CHEERING.

HOW'S THAT?

I SUGGEST YOU TAKE ANOTHER LOOK AT THE BOARD.

YOU FEEL YOU HAVE TIME TO MAKE CONVERSATION?

YOU CAN'T MAKE UP THE POINTS FROM HERE.

CUT OUT THE BIG TALK!

EH!

CHK

I'M JUST GETTING STARTED.

THE GAME ANALYSIS ON THE BIG BOARD WILL BE IN THIS 200-SEAT HALL...

EACH TEAM PARTICIPATING IN THE HOKUTO CUP WILL HAVE THEIR OWN WAITING ROOM... YES, HERE. WE'LL PROVIDE BEVERAGES.

THE PLAYERS AREN'T HERE FOR SIGHT-SEEING.

I UNDERSTAND.

THE HOTEL ROOMS FOR THE CHINESE AND KOREAN PLAYERS AND THEIR ENTOURAGES WILL BE...

PLEASE GIVE THEM QUIET ROOMS, NOTHING BY AN ELEVATOR OR THE LIKE.

NOW, ABOUT THE RECEPTION THE DAY BEFORE THE TOURNAMENT...

PLEASE LET US KNOW HOW ELSE WE CAN HELP.

THANK YOU VERY MUCH.

...OUR SOLE INTEREST IS IN SERVING ALL OUR GUESTS TO THE BEST OF OUR ABILITY.

I HAVE TO SAY...

I HOPE JAPAN DOES WELL, DON'T YOU?!

PRETTY EXCITING, ISN'T IT?

AGH!

YOU AND THAT MAN... JUST ANOTHER DAY AT THE OFFICE, EH?

OH... WELL...

WE TAKE NO POSITION CONCERNING TOURNAMENT RESULTS.

115

YOU ACTUALLY KNOW THE NAME OF ONE OF THE PLAYERS?

HMM...

GUESS IT'S UP TO ME TO GIVE A HOOT! GO AKIRA!

THERE'S ANOTHER.

...STRONG PLAYER ISN'T ENOUGH.

IT'S A TEAM TOURNAMENT, SO HAVING ONE...

THEY SAY AKIRA TOYA'S A CONTENDAH!

THE SECURITY GUARDS WERE CHATTING ABOUT IT.

BUT TOYA ASIDE, I HEAR OUR TALENT POOL'S VERY SHALLOW COMPARED TO THE CHINESE AND KOREANS.

MAYBE. HE SHOWS A LOT OF POTENTIAL, BUT HAS NO PROFESSIONAL RECORD TO SPEAK OF.

THEN JAPAN HAS A REAL SHOT!

HUH?!

116

...HAS SOME FASCINATING... ECCENTRICITIES.

HE JUST WENT PRO, BUT HIS AMATEUR RECORD...

THEN WHY'S HE GOT PEOPLE INTERESTED?

REALLY?

YOU'RE UP ON IT THAT MUCH?

OH...

IT'S WORKING INFORMATION, NOTHING MORE.

SOMEONE FROM THE KANSAI GO ASSOCIATION TOLD ME ON THE PHONE THE OTHER DAY.

YOU'LL NEED TO FUNCTION WITH DETACHMENT THE DAY OF THE TOURNAMENT.

CAREFUL... DON'T LET YOUR PATRIOTIC FERVOR GET OUT OF HAND.

FINE! BE THAT WAY!

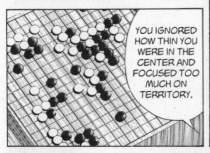

YOU IGNORED HOW THIN YOU WERE IN THE CENTER AND FOCUSED TOO MUCH ON TERRITORY.

BUT I SUPPOSE IT FELT GOOD WHILE IT LASTED.

HEH
HEH...

HEH...

I'M USED
TO BEING
TRAMPLED ON
BY YOUNGER
PLAYERS, BUT
TO BE...

KCHK

...BEATEN
SO BADLY
BY A
SHODAN?

I'VE
REALLY...
GONE
DOWNHILL.

I GUESS
THIS IS IT
FOR ME.

CLATTER

119

KCHNK

KCHNK

KCHNK

JUST MY LUCK, GOING UP AGAINST...

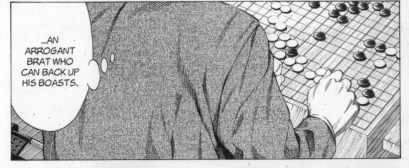

...AN ARROGANT BRAT WHO CAN BACK UP HIS BOASTS.

I LOSE.

THANK YOU FOR THE GAME.

THANK YOU VERY MUCH.

KCHNK

KCHNK

KLNK

...

KLNK
KLNK

YOU SEEM... TO KNOW A LOT ABOUT SHUSAKU.

DO YOU OFTEN REPLAY SHUSAKU'S GAMES?

KTNK

I BET YOU HAVEN'T STUDIED GO IN AGES.

YEAH, EVERY DAY.

KCHK

HMPH!

THNK

LOOKS LIKE SHINDO WON.

AND DON'T YOU DARE TRIP UP IN THE HOKUTO CUP QUALIFIERS.

COME AFTER ME, SHINDO.

BUT I'M NOT STANDING STILL EITHER.

Game 154 "Kenzan Ueshima!"

CLING
CLANG
CLING
CLANG

RATTLE
RATTLE

STAND!

BOW!

CHATTER
CHATTER
CHATTER

HEY!

TNK

TIME FOR ANOTHER TUTORING SESSION.

SKOOT

BAM

I DON'T HAVE TIME TO KILL.

WHAT'S IT TODAY? MATH? ENGLISH? EITHER WAY, HURRY UP AND GET IT OVER WITH.

...FAIL THE HIGH SCHOOL ENTRANCE EXAMS.

KEEP FOOLING AROUND AND YOU'LL...

DID YOU DO THE HOME-WORK I ASSIGNED?

HMPH!

WELL, GOOD FOR YOU.

FWMP

KANEKO IS HELPING MITANI STUDY AGAIN TODAY.

KANEKO ALREADY EARNED EARLY ADMISSION TO A TOP HIGH SCHOOL.

!

HEY, FUJISAKI! LOOKING FOR SHINDO?

SHE'S AMAZING. COMPARED TO HER, I'M...

SHINDO'S ABSENT TODAY.

OH! NO! NOTHING LIKE THAT! I...

UH-HUH...

WASN'T HE ABSENT THE OTHER DAY?

AGAIN?

ABSENT?

SHINDO'S MISSING A LOT OF SCHOOL THESE DAYS.

HERE. REDO THE ONES MARKED WITH AN "X."

MAKES SENSE I GUESS, WHAT WITH GO MATCHES AND STUDY SESSIONS...

GUESS HE DOESN'T PLAN TO GO TO HIGH SCHOOL.

WHAT KIND OF FUN WOULD HE HAVE AT SCHOOL ANYWAY? EVERYONE'S STRESSED TO THE MAX, CRAMMING FOR ENTRANCE EXAMS.

SHINDO LIVES NEAR YOU, DOESN'T HE?

SEEMS WE'RE ALL GOING OUR SEPARATE WAYS.

132

AKARI!

WHAT?! OH GOSH, I DIDN'T MEAN—

SO HE WON'T DISAPPEAR FROM **YOUR** LIFE.

AWRIGHT, SLAVE DRIVER... HERE!

FLAP FLAP

C'MON, WE GOTTA GET TO CRAM SCHOOL.

OH, SORRY! BE RIGHT THERE!

TSK... WHAT A MARTYR!

SCIENCE ROOM

TNK

SHEESH... WHY DO I NEED TO GET BETTER?

THE TOURNAMENT'S COMING UP SOON.

C'MON, OKAMURA! LET'S PLAY.

BOY, WHAT A SURPRISE WHEN WE FOUND OUT TSUTSUI... I MEAN, KAGA... WAS A MEMBER OF THE SHOGI CLUB.

I'M JUST HERE BECAUSE TSUTSUI... I MEAN, KAGA... ORDERED ME TO.

OH!

AFTER GETTING MORE INFO FROM THE UPPERCLASSMEN WE FINALLY FIGURED IT OUT...

HE WAS SO GOOD AT GO, I WAS SURE HE WAS A GO CLUB ALUM!

AND THE UPPER-CLASSMEN IN THE SHOGI CLUB ALL SAID, "DO **NOT** GO AGAINST KAGA'S ORDERS, WHATEVER THEY ARE!"

FLATTERY WILL GET YOU NOWHERE.

HMPH!

AND YOU'RE GETTING BETTER ALL THE TIME, OKAMURA. SO C'MON, LET'S PLAY!

THAT WAS A GREAT MOVE I—

YEAH, THAT'S TRUE!

IT'S NOT FLATTERY. THE OTHER DAY YOU PULLED OFF AN AMAZING SNAPBACK!

ARRR...

I THINK I'LL LEAVE NOW.

WAIT A SEC...

NOT AGAIN!

WHAT?!

BUT HOW WILL YOU STAY SHARP IF YOU DON'T PLAY?!

YOU CAN PLAY EACH OTHER.

YABE WILL BE HERE IN A MINUTE.

CLUB PRESI-DENT!

NOT YOUR WORRY.

AND DON'T GET SO WORKED UP ABOUT THIS TOUR-NAMENT. IT'S NOT THAT BIG A—

OKA-MURA!

THIS GUY SPRAINED HIS ANKLE AND HAD TO QUIT TRACK AND FIELD, SO HE'S INTERESTED IN JOINING THE GO CLUB!

HANG ON...

AND HE'S NOT A BEGINNER EITHER.

HEY...

FOR REAL?!

MY NAME'S UESHIMA. NICE TO MEET YOU.

HEY! WAIT!

HE SAYS HE PLAYS HIS DAD ALL THE TIME AT HOME.

IT'S ME, GOT IT?!

I'M THE THIRD MAN ON THE TEAM!

WE GO IN ORDER OF SKILL! DUH!

SHUT UP, OKAMURA!

THAT WE MIGHT HAVE A CHANCE IN THIS TOURNAMENT! I'M BREATHLESS WITH EXCITEMENT!

WHATTA YA MEAN BY THAT?!

WHOA THERE, YABE!

THAT'S NO PROB-LEM!

I ALREADY REGISTERED OKAMURA, YOU AND ME FOR THE TOURNA-MENT!

IT'S NOT THAT SIMPLE!

"OH, I SEE..." ?!

OH, I SEE...

UESHIMA HERE CAN PLAY UNDER THE NAME OKAMURA.

IT'S NOT?

UESHIMA! WE MIGHT BOTH BE FIRST-YEARS, BUT I HAVE SENIORITY IN THIS CLUB!

OKAMURA! WHAT'S THE BIG DEAL?!

OH... OH YEAH?!

FINE BY ME. I CAN BE "OKAMURA" OR YOU CAN... IT'S NO SKIN OFF MY NOSE.

ME, I WANNA GO INTO THIS TOURNA- MENT WITH A DECENT TEAM!

YOU HAVEN'T HELPED US GET NEW MEMBERS, AND YOU CAN'T PLAY GO WORTH A DARN! SO WHAT'S IT TO YOU?!

WELL, I'VE CERTAINLY PUT UP WITH A LOT.

...REALIZE YOU'D BECOME SO ENTHUSIASTIC!

I WANT HAZE TO MAKE A SHOWING AT THIS TOURNAMENT! A REAL SHOWING!

...GO ACTUALLY MEANS SOMETHING TO ME!

IN CASE YOU DIDN'T NOTICE...

AND MY FELLOW FIRST-YEAR DITCHES PRACTICE ALL THE TIME!

MY CLUB PRESIDENT'S A WORSE PLAYER THAN ME!

YABE, I DIDN'T...

WHOA! THE BETTER PLAYER SHOULD COMPETE, RIGHT?

SO TRUE!

YOU HAVE **ME** NOW, YABE.

IF I EXECUTE MY KILLER SNAPBACK, YOU WON'T KNOW WHAT—

UESHIMA! LET'S PLAY A GAME TO DECIDE IT!

LET'S PLAY! LET'S PLAY!

IT'S FINE WITH ME.

YOU'LL BE SLAUGH-TERED!

DECIDE WITH A GAME?

...AT HOME ARE MUCH HIGHER QUALITY.

THESE STONES LOOK CHEAP. OUR STONES...

KCHK

SO...THIS IS WHERE YOU PLAY? LOOKS FUN.

KLAK

AAAAH...

GULP

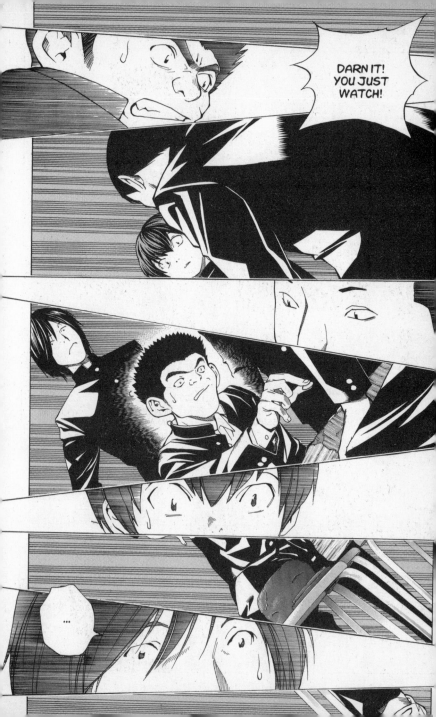

...A REALLY LOUSY PLAYER...

MOOOAN

UESHIMA, YOU'RE...

YABE!

YES! A LAND-SLIDE VICTORY!

I SEE. I GUESS IT'S POSSIBLE SOME PEOPLE PLAY GO FOR AGES AND NEVER GET BETTER.

NOT JUST 5-IN-A-ROW, BUT ACTUAL GO?

YOU PLAY OFTEN AGAINST YOUR DAD?

YEP!

YEP!

UNNH... I CAN'T BELIEVE THIS...

WE'RE OUT OF TIME. LET'S STOP HERE.

...

FIGHT ON!

HINOKI CRAM SCHOOL 2F

THERE'S STILL A MONTH BEFORE ENTRANCE EXAMS...

...SO DON'T WORRY. NOW, ALL TOGETHER...

FIGHT ON!

I'M STARVING.

SEE YA.

BYE.

144

IT'S FREEZ- ING.

SHIVER

I DUNNO... I CAN'T GET ANYTHING TO STAY IN MY BRAIN.

ONE MORE MONTH! WE CAN DO IT!

SEE YA.

HINOKI CRAM SCHOOL 2F

SHINDO LIVES NEAR YOU, DOESN'T HE?

SO HE WON'T DISAPPEAR FROM **YOUR** LIFE.

THE LIGHT'S ON IN HIKARU'S ROOM.

HE'S LUCKY... HE DOESN'T HAVE TO STUDY ANYMORE.

OH... HELLO, MRS. SHINDO!

AGH!

AKARI, RIGHT?

THE TEACHER SAYS IT'S BORDERLINE WHETHER OR NOT I'LL QUALIFY FOR MY TOP SCHOOL CHOICE.

ON YOUR WAY HOME FROM CRAM SCHOOL? IT'S HARD, ISN'T IT.

YES, MA'AM, IT SURE IS.

THANK YOU. UM... WERE YOU OUT SHOPPING THIS LATE?

THAT **IS** ROUGH. WELL, I WISH YOU THE BEST.

YOU DIDN'T ASK HIKARU TO GO OUT AND GET THEM FOR YOU?

...SPARE BULBS, SO IN SPITE OF THE COLD I POPPED OUT TO GET SOME.

THE BATHROOM LIGHT BURNED OUT, AND THERE WERE NO...

IT JUST DIDN'T SEEM RIGHT TO BOTHER HIM.

HE'S ALWAYS AT IT UNTIL VERY LATE.

WHICH IS PRETTY MUCH THE NORM THESE DAYS.

I COULD HEAR THE CLACK OF GO STONES IN HIS ROOM.

OH, THANKS. GOODNIGHT.

...GOODNIGHT. BE CAREFUL GETTING HOME.

WELL, THEN...

KLAK

TRRRRING♪ TRRRRRING♪

HUH?
UH...

MOM?

HELLO?

BEEP

I WAS... UH...
TALKING TO
KUMIKO SO I'M A
LITTLE LATE,
THAT'S ALL.

...DO MY
BEST.

I'LL...

...

YEAH...

YES, I'M
COMING
HOME.
I
KNOW...

BEEP

NOTHING,
OKAY?!

WHAT?
NO, IT WAS
NOTHING!

FIGHT
ON!

FIGHT
ON!

HIKARU NO GO

STORYBOARDS

 45

YUMI HOTTA

...IS ON EVERY WEDNESDAY EVENING AT 7:27 P.M. (REGIONAL AIR TIMES MAY VARY.)

THE HIKARU NO GO ANIME ON TV....

Oh, it's on now?

I'm watching right now.

...HAS CALLED ME JUST AT THAT TIME ON MORE THAN ONE OCCASION.

BUT MY EDITOR, TAKA-HASHI...

So about the storyboard corrections...

Well, I haven't seen it!

HE RE-PLIED, "I WATCHED THE VIDEO THEY SENT ME YES-TERDAY."

I ASKED, "YOU DON'T WATCH THE ANIME?"

Okay.

Please don't call me at 7:30 p.m. on Wednesdays.

SO ANOTHER EDITOR NAMED YOSHIDA GOT ASSIGNED TO HIKARU NO GO.

THAT'S TAKAHASHI FOR YOU, BUT HE GOT PRO-MOTED TO ASSISTANT EDITOR-IN-CHIEF.

Game
155
"The
Two
No-
Shows"

CLICK

KLAK

AGH!

GOOD! STOP RIGHT THERE, PLEASE.

DON'T GET STIFF JUST BECAUSE YOU'RE THE TOP BATTER IN THE SHINSHODAN TOURNAMENT!

RELAX... RELAX...

CLICK

JAPAN GO ASSOCIATION

WHAT'S THE MATTER, EH? YOU SEEM TENSE.

WHAT'S THE MATTER? THOUGHT ALL YOU YOUNG FOLKS LIKED TO GET YOUR PICTURE TAKEN.

CLICK

PLEASE, SIR, I...

AFTER ALL, YOU'RE THE PROMISING YOUNG PRO WHO PASSED THE ENTIRE PRO TEST **UNDEFEATED!**

I SUPPOSE IF I PLAY YOU WHEN YOU'RE A LITTLE TENSE, YOU'LL JUST TOSS ME OFF LIKE IT'S NOTHING.

HUH? WELL... ER... I DON'T...

TOO BAD. A GO PLAYER NEEDS PHYSICAL STAMINA. COME ALONG WHEN I PLAY GOLF AND BE MY CADDY. THAT'LL HELP YOU GET IN SHAPE.

WHEW! KUWABARA SENSEI'S PUTTING THE PRESSURE ON ALREADY.

KUWA-BARA'S IN CONTROL.

CLICK

ER... N-NO...

YOU'RE A BIT THIN, BUT OTHERWISE WELL SET UP. DO YOU PLAY ANY SPORTS?

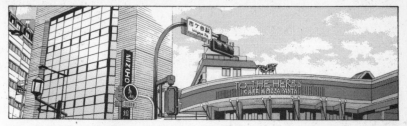

CAN'T WAIT TO SEE ISUMI PLAY KUWABARA IN THE SHINSHODAN TOURNAMENT.

IT OUGHTA BE SOME GAME.

THAT'S...

!

HEYYY... SHINDO!

HI, KADOWAKI.

YOU HERE TO WATCH ISUMI'S GAME TOO?

OH...

YOU KNOW MY NAME? I'M IM-PRESSED.

AH!

YOU REMEMBER THAT? I'M HONORED.

SURE DID.

SAW ME IN *GO WEEKLY*, PERHAPS?

I WAS SURPRISED WHEN I SAW THE PHOTOS OF THE NEW PLAYERS WHO PASSED THE PRO EXAM THIS YEAR. I REMEMBERED PLAYING YOU ONCE WHEN I WAS AN INSEI.

THAT DAY, I ASKED YOU FOR A GAME AS A KIND OF WARM-UP BEFORE THE EXAM.

YEAH?

I'D PLANNED TO TAKE THE PRO EXAM THE SAME YEAR AS YOU.

I THOUGHT, "I'M NOT GOOD ENOUGH YET."

I REALIZED HOW UNREADY I WAS, AND POSTPONED FOR A YEAR.

BUT EVEN THOUGH YOU WERE STILL AN INSEI, YOU CLOBBERED ME.

I REALIZED I HAD TO AIM HIGHER...

...AND REALLY GET MY BUTT IN GEAR.

I MAY HAVE HAD THE CONFIDENCE TO PASS, BUT NOT THE CONFIDENCE TO GO UP AGAINST PLAYERS LIKE YOU.

GEE...

THAT'S HOW GOOD YOU WERE.

I'M A CONFIDENT GUY, BUT YOU MADE ME THINK.

BUT THEN YOU REALLY HIT THE WALL, LOSING THREE TIMES IN THE PRO EXAM.

SO WHAT IS IT?

AND CAVED COMPLETELY IN YOUR SHINSHODAN GAME AGAINST KOYO TOYA.

...

EVEN AFTER MAKING PRO, YOU HAD A STRING OF FORFEITS.

WHAT'S THE DEAL WITH YOU?

HE SAID HE'D COME SEE ISUMI'S SHINSHODAN GAME.

WHAT'S KEEPING SHINDO?

HONDA, YOU LOST SIX IN THE PRO EXAM, RIGHT?

SO DID KADO-WAKI.

KLAK

THIS YEAR THOSE TWO TOOK ALL THE WHITE STARS! IT WAS A VICIOUS BATTLE FOR THE REST OF US!

ISUMI WON ALL OF HIS GAMES, AND KADOWAKI ONLY LOST HIS GAME AGAINST ISUMI.

I'M SURPRISED YOU PASSED WITH A RECORD LIKE THAT.

GCHK

THE JAPAN-CHINA-KOREA
JUNIOR TEAM TOURNAMENT

THE HOKUTO CUP

The Japan-China-Korea   The Hokuto Cup
$4.95

!

STILL NOT HERE...

KLAK

IF IT'S BASED ON WIN RECORDS, I'M BOUND TO GET ONE OF THE REMAINING SPOTS ON JAPAN'S TEAM!

HEY, I HEAR TOYA WAS PRE-SELECTED TO REPRESENT JAPAN.

SLAM

HEH! YOU LOST TO ME!

I WASN'T PLAYING ALL OUT THAT GAME!

I GOT TO THE SECOND ROUND OF THE HON'INBO PRELIMS!

...THERE'S SOMEONE PROMISING IN THE KANSAI GO ASSOCIATION.

SPEAKING OF THE NEW PROS, I HEAR...

...PLAYERS 18 AND UNDER USUALLY DON'T HAVE LONG TRACK RECORDS, SO IT'D BE HARD TO JUDGE BY THAT.

SHINDO MADE IT THAT FAR TOO. ANYWAY...

KrrK

...THE TOURNAMENT'S EVEN OPEN TO NEW PROS LIKE YOU, HONDA.

TRUE. STILL...

MAYBE WE'LL ALL BE IN ONE QUALIFYING TOURNAMENT!

AND JAPAN GO ASSOCIATION PLAYERS FROM THE CENTRAL AND WESTERN AREAS.

OH YEAH? WHO?

THAT'S RIGHT, THOSE GUYS CAN PARTICIPATE TOO.

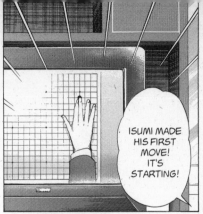

ISUMI MADE HIS FIRST MOVE! IT'S STARTING!

THAT WOULD BE PRETTY COOL.

I WONDER IF ISUMI'S OKAY, OR IF HE'S FEELING ANXIOUS.

KSHH

KSHH

...BUT THAT LOOKED LIKE A PRETTY CONFIDENT FIRST MOVE.

HE'S PLAYING KUWABARA SENSEI...

KLAK

REMEMBER SHINDO'S GAME LAST YEAR?

HE TOOK 20 MINUTES TO PUT HIS FIRST STONE DOWN.

NAH... THEY'VE NEVER MET.

MAYBE THEY'RE HANGING TOGETHER.

AND WHERE'S KADOWAKI?

I WONDER WHAT'S KEEPING HIM?

I HEARD KADOWAKI MUTTERING SOMETHING... NOW I GET IT! HE WAS TALKING ABOUT SHINDO!

WHAT'S UP, HONDA?

WHAT'D YOU SAY...?

...

THAT'S IT! IT WAS SHINDO!

HUH? OH, I SAID THEY'D NEVER **MET**.

OSHIMA, FUKUI, HONDA...

163

IT'S HARD TO BELIEVE...

THOSE WERE THE GUYS HE LOST TO LAST YEAR...

AND ONCE I'D PASSED, I WAS SO HAPPY I FORGOT ALL ABOUT IT.

...BUT I WAS ON THE BRINK OF PASS OR FAIL ON THE EXAM, SO I DIDN'T PAY MUCH ATTENTION.

I HEARD MY NAME MENTIONED...

IT WAS AT LUNCH DURING THE LAST MATCH OF THE PRO EXAM.

YEAH, THOSE WERE THE ONES WHO BEAT SHINDO IN LAST YEAR'S PRO EXAM.

OSHIMA, FUKUI AND HONDA...

SHINDO, SHINDO, SHINDO... GEEZ...

HUH? WHAT OTHER THING?

MAYBE THAT OTHER THING WAS ABOUT SHINDO TOO!

GOOD MEMORY, WAYA.

KCHK

OH!

THAT'S CUZ HONDA AND I WERE NECK AND NECK WITH SHINDO RIGHT UP TO THE END.

...BUT ACCORDING TO THE RUMOR...

OLD NEWS, I KNOW...

THERE WAS THIS ONLINE RUMOR THAT KADOWAKI PUT OFF TAKING THE PRO EXAM FOR A YEAR.

HE WAS EMBARRASSED AND DECIDED HE NEEDED A LOT MORE STUDY AND TRAINING.

...THE REASON WAS BECAUSE HE'D LOST TO SOME KID.

I'D BE MONEY ON IT! KADOWAKI LISTED THE GUYS SHINDO LOST TO IN THE PRO EXAM, AND HONDA SAYS HE SEEMED PRETTY DISAPPOINTED.

SO MAYBE THAT KID WAS SHINDO?

AT THE TIME I COULDN'T THINK OF ANY KID WHO COULD PUT ANYONE OFF LIKE THAT. I JUST FIGURED KADOWAKI WASN'T VERY GOOD.

IN OTHER WORDS, HE COULDN'T BELIEVE SHINDO LOST TO THE LIKES OF THOSE THREE.

NEITHER OF THEM IS HERE. THAT MUST MEAN SOMETHING.

WHAT'S THE POINT OF OBSESSING ABOUT SHINDO?

OSHIMA AND FUKUI, FINE! BUT ARE YOU GONNA LUMP ME IN THERE? ME?

EXACTLY!

FORGET SHINDO!

DON'T BE SILLY! THE SURE THING IS ME!

SO SHINDO'S SURE TO MAKE THE HOKUTO CUP TEAM?

CLICK

I WAS JUST SAYING THAT IT'S NOT...

SHINDO! OCHI WAS JUST...

OH!

OH, IT'S YOU, SAKURANO.

ISUMI KNOWS SAKURANO BECAUSE OF KYUSEIKAI?

I DIDN'T MEAN IT LIKE THAT!

YOINK

GREAT TO SEE YOU TOO, WAYA!

I THINK SO...

GRIN

SO WHAT'S THE STORY?

THOSE THREE GAMES YOU LOST...

...WAS IT BECAUSE YOU'RE A KID AND PLAYED UNEVENLY?

SERIOUSLY, IT'S THE ONLY EXPLANATION I CAN THINK OF.

BECAUSE, SERIOUSLY AGAIN, YOU SENT ME OFF WITH MY TAIL BETWEEN MY LEGS. FOR A WHOLE YEAR.

PRETTY FUNNY, HUH?

NO.

BUT NOW I WONDER... WHAT REALLY HAPPENED?

I'D PINNED MY HOPES AND EXPECTATIONS ON YOU.

SEEMS LIKE A DREAM OR AN ILLUSION NOW...

CAN I?

LOOK, YOU'VE BEEN PLAYING WELL LATELY. CAN I STILL HAVE HIGH HOPES FOR YOU?

169

...BEFORE ISUMI'S SHIN-SHODAN GAME IS LAST WEEK'S NEWS.

LET'S GET IN THERE...

BUT HEY, I'VE HELD US UP LONG ENOUGH.

KADOWAKI...

DO YOU WANT TO PLAY A GAME WITH ME?

RIGHT NOW?

A WORD ABOUT HIKARU NO GO

IT'S GENERALLY HELD THAT WOMEN GO PLAYERS
ARE NOT AS STRONG AS THEIR MALE COUNTER-
PARTS. BECAUSE OF THAT, IN THE REAL LIFE GO
WORLD THERE ARE PRO EXAMS AND TOURNAMENTS
JUST FOR WOMEN.

BUT—

IN 2000, A WOMAN IN KOREA TOOK THE TITLE IN A
GENERAL GO TOURNAMENT FOR THE FIRST TIME.
AND IN 2002 A KOREAN WOMAN BARELY 18 YEARS
OLD BEAT A JAPANESE MASTER AT AN INTERNA-
TIONAL TOURNAMENT.

THE RISE OF WOMEN

IT'S ONE OF THE EXCITING THINGS TO LOOK
FORWARD TO IN THE WORLD OF GO.

I WONDER WHEN THEY'LL FINISH THE REMODELING WORK ON THE FIRST FLOOR.

Game 156 "Hikaru vs. Kadowaki"

THE SHOP FROM THE FIRST FLOOR'S HERE TOO.

SURE.

WE'RE GONNA PLAY A GAME HERE, OKAY?

GO GAME ROOM

THIS MATCH ROOM ON THE SECOND FLOOR LOOKS A LITTLE DIFFERENT THAN IT DID A YEAR AND A HALF AGO.

HERE'S A GOOD SPOT. WE WON'T BE DISTURBED.

SHINDO?

WHAT'S THE MATTER?

出下さい

THIS'LL BE TOUGH...

SKOOT

WELL, IT'S A YEAR AND A HALF LATER.

WHY?

HIKARU - no GO ↰ !!

Hikaru   Akira
   Sai   Akari

... READY TO MEASURE EACH OTHER NOW AGAINST WHAT WE WERE THEN!

I KNOW. AND HERE WE ARE, ACROSS A BOARD...

IT WAS YOUR ONLY LOSS.

ISUMI! WELL, YES, I DID LOSE TO **HIM**, BUT...

KTNK

...BUT I BET YOU'VE READ ABOUT IT IN *GO WEEKLY*.

I'D LOVE TO BRAG ABOUT HOW I DID IN THE PRO EXAM...

YOU LOST TO ISUMI.

I HAVE.

ONE THING I LOOK FORWARD TO AS A PRO IS THE CHANCE TO AVENGE MYSELF!

DOESN'T MEAN HE'S BETTER'N ME, Y'KNOW!

...

KSHH

...IT HAPPENED QUICKER THAN I EXPECTED.

ANOTHER IS GETTING TO PLAY YOU AGAIN. GOTTA ADMIT...

KCHNK

KCHNK

CLATTER

KTNK

CLATTER

ONEGAI-SHIMASU.

KLAK

A STAR POINT, EH?

KLAK

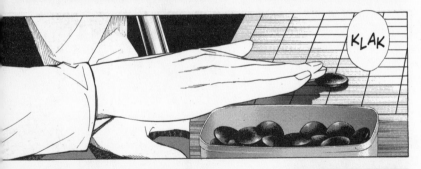

KLAK

KLAK

ANOTHER STAR POINT... TWO IN A ROW.

KLAK

...A YEAR AND A HALF AGO?

INTERFERE EARLY ON, JUST LIKE I DID...

THREE STAR POINTS IN A ROW. HMM...

HOW SHOULD I PLAY THIS?

THAT TIME HE WAS TOYING WITH ME, BUT NOW I WON'T LET THAT HAPPEN.

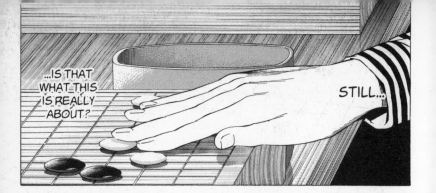

...IS THAT WHAT THIS IS REALLY ABOUT?

STILL...

AND... CAN I POSSIBLY FOLLOW HIM?

...BECAUSE I WANT TO SEE DOWN THE PATH SHINDO'S TAKING.

SURE, I WANT TO WIN AS MUCH AS ANYBODY, BUT THIS GAME, FOR ME, IS ALSO A JOURNEY OF DISCOVERY...

THAT'LL BE 1050 YEN.

THAT'S SHINDO SHODAN, AND I BELIEVE THE OTHER IS A NEW SHODAN NAMED KADOWAKI.

THOSE TWO LOOK KINDA FAMILIAR.

HE WAS A BIG DEAL IN UNIVERSITY TOURNAMENTS!

OH, KADOWAKI!

182

I THINK I'LL GO TAKE A LOOK.

THANK YOU, SIR.

...

KLAK

I'LL LEAVE 'EM ALONE.

THEY'RE... INTENSE!

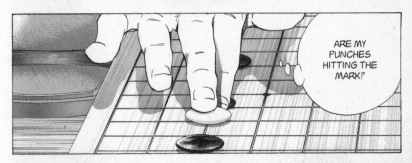

ARE MY PUNCHES HITTING THE MARK?

THEY'RE HITTING THE MARK, RIGHT?

KLAK

OR IS HE DODGING THEM?

BUT THERE'S NO SIGN OF A COUNTER-ATTACK.

KLAK

IS HE EVEN THINKING OF A RESPONSE?

KLAK

KLAK

HIS GO IS NOT LIKE IT WAS...

!

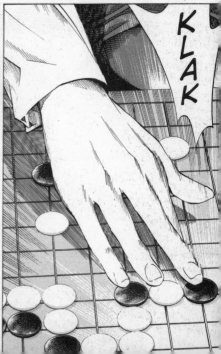

KLAK

SO
NOW IT
STARTS!

AH...

HE'S CLAMPED MY HANE!

KLAK

KLAK

!

HMPH!

...

WHAT DID YOU SAY YOUR NAME WAS?

SHIN- ICHIRO ISUMI.

HIS NAME IS ISUMI.

KOFF

189

...BUT EVER SINCE TOYA'S BOY WENT PRO THERE'S BEEN AN UPTICK IN QUALITY. MODEST, PERHAPS, BUT WORTH NOTING.

NEW PROS HAVEN'T AMOUNTED TO MUCH IN RECENT YEARS...

ISUMI, HUH?

I GUESS I'LL HAVE TO REMEMBER IT.

ESPECIALLY WITH THAT ONE...

...WHOSE TRUE ABILITIES HAVE YET TO BE TESTED.

ALMOST MAKES ME THINK IT'S WORTHWHILE TO HAVE STUCK AROUND FOR SO LONG.

KLAK

!

KCHK

HMM...

KSHH

...

...

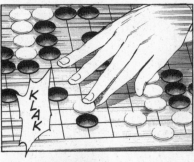

The End of One Step Forward!

Preliminary matches get underway to determine the four players from Tokyo who will vie for a spot on the Japanese team that will compete in the Hokuto Cup. Hikaru aims to secure a position on the team but gets his head handed to him when he finally goes up against a seasoned pro in an official tournament game. Can he rally from this with the competition heating up on practically all fronts?

COMING AUGUST 2010